LEE'S ESSENTIALS OF WIRELESS COMMUNICATIONS

LEE'S ESSENTIALS OF WIRELESS COMMUNICATIONS

WILLIAM C.Y. LEE, PH. D.

Chairman, LinkAir Communications, Inc.
(formerly Vice President and Chief Scientist,
Vodafone AirTouch PLC)

McGraw-Hill
New York • San Francisco • Washington, D.C. • Auckland
Bogotá • Caracas • Lisbon • London • Madrid • Mexico City
Milan • Montreal • New Delhi • San Juan • Singapore
Sydney • Tokyo • Toronto

Cataloging-in-Publication Data is on file with the Library of Congress

McGraw-Hill

A Division of The McGraw·Hill Companies

1 2 3 4 5 6 7 8 9 0   DOC/DOC   0 9 8 7 6 5 4 3 2 1 0

ISBN 0-07-134542-6

The sponsoring editor for this book was Steve Chapman, the editing supervisor was Steven Melvin, and the production supervisor was Sherri Souffrance. It was set in Fairfield per the NBF specs by Joanne Morbit of McGraw-Hill's Professional Book Group composition unit, Hightstown, N.J.

Printed and bound by R. R. Donnelley & Sons Company.

This book is printed on recycled, acid-free paper containing a minimum of 50% recycled, de-inked fiber.

McGraw-Hill books are available at special quantity discounts to use as premiums and sales promotions, or for use in corporate training programs. For more information, please write to the Director of Special Sales, Professional Publishing, McGraw-Hill, Two Penn Plaza, New York, NY 10121-2298. Or contact your local bookstore.

CONTENTS

CHAPTER FIVE. LEARN FROM THE PAST   119

CHAPTER SIX. APPLICATION OF CDMA   169

CHAPTER SEVEN. WHAT IS OUR FUTURE?   197

CHAPTER EIGHT. INTERNET AND WIRELESS FUTURE   241

PREFACE

There have been many changes in the wireless communications industry in the past 15 years. In this book I will try to describe the latest wireless communications systems and technologies, some of them have prevailed and some did not. I have to say that the fate of each one depended on "effort" plus "luck." Effort can be controlled, but luck, no. Speaking of luck, I will tell you how I got into the wireless communications field.

In December 1963, when I finished my doctoral dissertation at Ohio State University on topics related to space communication, I got an offer from Bell Laboratories to work in the new field of satellite communication. By the time I reported to work on March 1, 1964, the job had been eliminated; the U.S. Congress had passed a bill to form a company to run the satellite communications business, called Communication Satellite Company or ComSat for short. I could either go to Washington, D.C. to join ComSat or be assigned a different job. When I found out that the new job was in mobile communication, I told them frankly that I did not study mobile communication in school. Their reply was that "nobody knows the field." The Mobile Communication Research Department had just been formed. I was the first person hired from the outside. I did not choose mobile communication, but I was lucky to become a part of it.

Tracing over the 15 years of my mobile communications work at Bell Labs, I started in the Research Division, moved to the System Division, then to the Switching Division, and finally to the Development Division. With this fruitful experience and knowledge, Bell Laboratories elected me to teach a course, "Mobile Communication Theory." In 1978 this was a Bell Laboratories in-house course. Many well-known scientists at Bell Lab attended the course. The material from the lectures inspired me to write my first book, *Mobile Communication Engineering*, published by McGraw-Hill in 1982.

AT&T had been trying to acquire the 800-MHz cellular license from the FCC since 1974. The reason they never succeeded was the Radio Common Carrier (RCC) Association, afraid that AT&T would monopolize the mobile communication industry, had asked the FCC to delay the issuance of the 800-MHz license. By 1979, I felt I should not wait at Bell Labs for the license to be granted. I joined ITT DCD (Defense Communication Division), working on the military mobile communication project. I was granted two patents in spread spectrum (SS) modulation. One was for anti-jamming and the other for covert communication. In 1984, using artificial intelligence (AI), I invented connnectionless (without a master station) communication to be used in battle situations. It took the U.S. Patent Office Examiner 3 years to grant me a patent because the field was new to him.

In 1981, 20 MHz of the 800-MHz cellular spectrum was divided by the FCC into two bands, Band A and Band B. Band A licensed non-telephone companies (paging or dispatching). This was called the non-wireline band. Band B licensed telephone companies and was the called the wireline band. In 1984, most of the regional Bell Operating Companiess (RBOC, aka the baby Bells) started deploying cellular systems. In 1985, PacTel Celular (a subsidiary of Pacific Telesis Company) asked me to join their company. At that time, some companies were lobbying the FCC to have a new SSB (single side band) system in cellular. A 30-kHz FM channel could be replaced by six SSB channels. Thus, they claimed the SSB system would have more capacity and a higher spectrum efficiency than the FM system.

On August 2, 1985, I was invited by the FCC to talk on the comparison of spectrum efficiency between the SSB and FM systems. My finding was that, in order to maintain the same voice quality, the requirement of the co-channel reuse distance for SSB was larger that that for FM in a cellular mobile system. As a result, there was no difference between the two systems in spectrum efficiency. After my talk SSB promoters became quiet. Can you imagine! There would have been two analog systems, FM and SSB. It would definitely have slowed down the growth of the cellular industry at the start-up stage.

In 1987, cellular system capacity started to become an issue. We concluded that no analog system would increase capacity and that we needed to go to digital systems for greater capacity.

On September 3, 1987, the FCC invited representatives of three large manufacturing companies and me from an operating company to discuss future cellular systems. I presented a new formula that could evaluate the capacity of each digital system. This formula was later used by industry to compare different systems.

I was honored to be chosen by the cellular industry to be a co-chair of the CTIA Advanced Radio Technology Subcommittee (ARTS). At that time, I believed that FDMA was a low-risk choice and that it could be deployed in 1990. ARTS sent a letter to four companies—AT&T, Motorola, Northern Telecom, and Ericsson—and asked them if they saw any difficulty in choosing FDMA. The answers were, "no." By that time, I resigned from the co-chair and started developing my patented microcell system. In 1988, after much discussion and debate in the industry, TDMA was chosen. I was disappointed. In 1989, Qualcomm excutives and key engineers came to see me to discuss the possibility of using CDMA. In February 1989, I pointed out to them that power control was needed in a terrestrial CDMA system because of near-far interference. In April 1989, Qualcomm discovered the technology of implementing power control. That November Qualcomm's CDMA demonstration shocked the global cellular industry. In 1990, Dr. Han-Su Park from Korea asked me to give a 3-day seminar in Seoul. In the seminar, when I briefly described the new CDMA system, I drew the attention of ETRI (Electronic Technology and Research Institute). Then, Alan Salmasi, from Qualcomm, and I went to Korea to give a CDMA seminar, which opened the doors for the Korean government to pursue CDMA technology. We claimed that the capacity of CDMA could be 20 times larger than AMPS. However, the American CDMA system would not realize its capability until the Korean CDMA market reached 1 million subscribers in September 1996. CDMA technology had proven its value.

From 1990 to 1996 many other systems were developed—GSM, CDPD, DECT, NAMPS, PDC, PHS, and iDEN. In this book, I review each system. I also chose to review a third-generation (3G) system with CDMA technology. I have had the opportunity to be involved in the harmonization process among the different proposed 3G systems through the Operator Harmonization Group (OHG).

Since the Internet is growing quickly, the mobile Internet will become our future trend. As a result, the future mobile radio network will be the wireless IP core network. It was with much pleasure that AirTouch was able to work with three companies—Cisco, Hyundai, and Telos—to put up a total-solution IP core network demonstration. On December 15, 1999, the demonstration was held in Reno, Nevada, and was very successful. It was the first demonstration of the feasibility of an IP core network without switches. We were all proud of our result.

In 1998, McGraw-Hill Executive Editor, Steve Chapman asked me to write a book about the past, present, and future of mobile communication based on my involvement. Since what I write here is viewed from my perspective, the reader should understand that there might be some biases. Thirty-four years of my observation and involvement—1964 to 2000—is written about in this book. I hope I have given the reader something of value.

William C.Y. Lee, Ph. D.

ACKNOWLEDGMENTS

The idea of having me write this book came from Steve Chapman, executive editor of McGraw Hill Company. At first, he wanted the book title to be *Lee's Wireless Communication*. I felt it was a bit too much for me. Then the title *Lee's Essentials of Wireless Communication* was agreed upon. All his efforts deserve special mention.

This book took more time than I had originally estimated. For the last two yews, I have spent my time in leading two exciting projects. One was the successful low-cost infrastructure equipment trial, held in Modesto, California, sponsored by AirTouch with Samsung, Hyundai, Telos, Teco, and Celletra. The other project was the total solution of wireless IP core network held at Reno, Nevada, sponsored by AirTouch with Cisco, Hyundai, and Telos. I had planned to include these two projects in this book, but after the manuscript was sent out I realized that the Modesto trial's material was left in a drawer. I feel very badly about this omission. However, I should like to deeply thank all the executives and the outstanding engineers from the different companies who accomplished these projects and provided me with more new knowledge to include in this book

During my writing, Sam Ginn and Arun Sarin of AirTouch gave me a great deal of encouragement. In Chinese the proverb says, "Every time you drink water remember where the source is." I would also like to thank my two advisors from Ohio State University, Professor W. R. Walters and Professor Leon Peters, as well as my two mentors at Bell Labs, Dr. C.C. Culter and Dr. Frank Blecker. From my 15-year career with PacTel, which became AirTouch, then Vodafone-AirTouch, then finally Vodafone-US, I think all my colleagues who helped me, especially my assistant, Mrs. Carla Sherbert. I owe all of them my deepest appreciation.

Now, I am assuming a new position with LinkAir Communications, a company that holds a new coding technology, which can strongly enhance the FDD system and will be a breakthrough technology for TDD. I hope that the industry will give LinkAir constructive advice and not destructive comments. I would be most appreciative. I believe that we can all work together for our future in the information age.

Last, but not least, I have to thank my wife, Margaret, for providing me the time to finish this book. She takes care of my health and encourages my work with her sweet compliments. I have already promised her that I will only write books with co-authors from now on. She is my inspiration.

HOW THE TELEPHONES, WIRELINE, AND WIRELESS WERE BORN

1.1   THE SUCCESS OF TELEPHONES

The use of the telephone is growing in our daily life. But imagine what it was like for people 120 years ago, when people first started using this odd-looking device to communicate with their friends and neighbors. Before the telephone was invented, talking to one another meant having face-to-face conversations, not listening to each other's voice as it traveled back and forth through a line. People had to adjust to this new way of communicating with each other, so it is not surprising that it took some time for them to get comfortable with using the telephone. However, it wasn't too long before people began to depend on their telephones—perhaps even wondering how they ever got along without them. A telephone conversation is still the next best thing to a face-to-face conversation. Today's teenagers would certainly agree—they're on the phone for hours at a time. If a household went without a phone for just one day, every family member would be inconvenienced.

The success of telephony since 1876 has been based on several deciding factors. How did the telephone become so popular throughout the world today?

1.1.1 TIMING

Alexander Graham Bell and Elisha Gray filed their telephone patent applications on the same day, but Bell filed his patent a few hours earlier. This historical fact made one person famous and the other an unknown.

On March 10, 1876, Alexander Graham Bell succeeded in speaking over a telephone wire with his assistant Thomas W. Watson. We have all heard this story. However, Elisha Gray from Western Union was also working on the telephone at the same time. Bell filed his first patent on February 14, 1876, and Gray filed his patent a few hours after Bell. The United States Patent Office issued Bell's patent on March 7, 1876. In September 1878, the Bell Company sued Western Union to protect Bell's telephone patents. During 1879, the Bell Company filed more than 600 lawsuits against Western Union over Bell's patents. Finally, at the end of the same year, Western Union acknowledged Bell's patents and agreed to stay out of the telephone business. If Gray had filed his patent a few hours earlier than Bell, what would history look like today?

1.1.2 Strategy

Bell's father-in-law, Gardiner G. Hubbard, was a lawyer and had suggested leasing the telephone units to the subscribers instead of selling them. This decision gave the Bell system more freedom to improve and upgrade the telephone system as technology developed.

1.1.3 Government Regulation

In the United States there were about 2000 independent telephone companies besides the Bell System. The U.S. government imposed backward compatibility rules to apply to all the existing phones whenever any new phone technology was deployed. Today the old rotary phones still work.

The three factors described above, timing, strategy, and government regulations, were implemented in the telephone industry's early years and contributed to its success today. Of course, this leasing strategy may not be applied to today's markets. For example, in the wireless communication industry, the new technologies advance rapidly and the old technologies retire quickly. Also phones are very inexpensive. The leasing strategy of phone unit services no longer has advantages.

1.2 HISTORY OF THE HIGH-CAPACITY-SYSTEM STUDY

In 1947, in Federal Communications Commission (FCC) Docket 8658, the Bell System proposed a broadband urban mobile system and requested a 40-MHz band in the region between 100 and 450 MHz. This proposal was made at about the time that the first mobile telephone system, at 150 MHz, was installed in St. Louis; it illustrates the fact that the current systems, at 35, 150, and 450 MHz, have been intended, from their inception, to be stop-gap service offerings to keep the Bell System in the mobile business until a more appropriate service could be offered.

The outcome of FCC Docket 8658 was that the request was denied, not because the FCC considered the proposed service to be undesirable, but because no such band was available for allocation. In 1949, the FCC initiated Docket 8976, which considered the allocation of the UHF band from 470 to 890 MHz. The FCC entertained the possibility of allocating the lowest 30 MHz of this band to common carrier mobile radio operation but ultimately allocated the entire band to broadcast television. As in 1947, it found the proposed service too desirable, stating,

> In arriving at this conclusion we are forced to resolve a conflict between two socially valuable services for the precious spectrum space involved. We find that the needs of the two are compelling.

In denying the Bell System petition, the FCC hinted at the Dockets to come, stating,

> But while we find and conclude that there is, on the part of the common carrier mobile service, the need for further expansion of service beyond that already provided by our rules and regulations and by techniques now being employed, we do not conclude that the only available solution to the common carrier land mobile service lies in the utilization of the frequency band 470–500 MCS.

In 1958, in Docket 11997, AT&T once again proposed a broadband mobile telephone system, requesting a 75-MHz band from 764 to 840 MHz. This was based on earlier studies at Bell Labs that had indicated that operation at 800 MHz would be feasible. After hearing testimony, the FCC took no action. It did not deny the petition, but neither did it allocate the spectrum. In retrospect, however, it appears that the AT&T filings and testimony in Docket 11997 provided the genesis for Docket 18692. On July 26, 1968, the FCC released a Notice of Inquiry and Notice of Proposed Rule Making in Docket 18262. The FCC proposed to allocate 40 MHz for private mobile radio systems and 75 MHz for common carrier high-capacity mobile radio systems. It is the latter allocation to which this book has been addressed. Not only are the two specific allocations strikingly similar, but, in their Notice of Inquiry and Notice of Proposed Rule Making in Docket 18262, the FCC repeatedly makes reference to the 1958 AT&T testimony.

In the years between 1958 and 1968, Bell Laboratories and AT&T continued to study, at a low level of effort, the systems and equipment aspects of a high-capacity mobile radio service. None of these studies, including the last one in 1965, were optimistic about the economic viability of such a service. Therefore, in 1968, when the FCC reopened the question in Docket 18262, the response by AT&T was a cautious one. In its filing of February 3, 1969, AT&T proposed to carry out a two-phase program. Phase I would include the studies and exploratory development work required to determine the characteristics and feasibility of high-capacity mobile systems and the suitability of the Commission's specific frequency allocation plan for common carrier use. This phase was to begin when there was "reasonable assurance" that the 75 MHz would be made available to common carriers for high-capacity mobile radio service and was to be completed in 18 months. Phase II of the Bell System program was to be undertaken only if the results of Phase I justified further work and if the Commission concurred.

On May 20, 1969, the FCC issued its First Report and Order and Second Notice of Inquiry. This response differed from the Original Notice of Inquiry in three ways. It moved the proposed 75 MHz slightly, to a more advantageous continuous block, 806 to 881 MHz; it "set aside" the frequencies, which is apparently a stronger commitment than the "reasonable assurance" AT&T requested; and it excluded all but the wire-line common carriers from the band. This last exclusion was the subject of various Petitions to Reconsider, and on July 30, 1971, in a Second Memorandum Opinion and Order, the Commission deleted it. With the May 20, 1970, Report and Order, the Phase I study officially began.

The 18-month Phase I study ended in November 1971, and AT&T reported its results to the FCC at that time.[1]

1.3   THE BIRTH OF THE CELLULAR SYSTEM[2-5]

The concept of reusing the same frequency in cellular systems, called frequency reuse, was first proposed by Doug Ring at Bell Labs in 1957. Then W. D. Lewis planned a broadband mobile phone system in 1960.[6] Later, C. C. Cutler created a new mobile radio department in 1964 and assigned W. Jakes to lead the research with a group of engineers in the Labs between 1964 and 1972.

From 1970 to 1974, AT&T continuously asked the FCC to allocate a spectrum for cellular service. Because in the late 1960s most UHF spectrums were allocated to the TV industry from Channel 2 (54–64 MHz) to Channel 83 (884–890 MHz); also the fixed wireless (point-to-point) and aeronautical fixed wireless had been allocated in general from 1.6 to 30 GHz. There was no spectrum that could be used by cellular service. The FCC did not make any decision or give any hint to AT&T that it might. AT&T, on the other hand, negotiated with the FCC and tested the spectrum at 800 MHz, which was the high end of TV channels. The FCC informed AT&T that the chance of allocating 800 MHz was not good. Even Sam McCannery, Branch Chief of FCC, jokingly said that Bell Labs had invented many things so why not invent its own spectrum. At that time, Bell Labs also tested the system at 10^{-7} and 60-GHz bands.[5,8] The 10-GHz band was not used by the U.S. government and the 60-GHz band, which was a high attenuated band, could be allocated to any user. Thanks to cable TV's deployment, broadcast TV did not use all 82 channels in 1974. The FCC reallocated the spectrum in 800 MHz from the broadcast TV channels (Channels 73 to 83) to cellular operators.

In the 1970s, AT&T was developing the first-generation cellular phone system in Holmdel, N.J., and requested the FCC Docket 18262 to allocate a segment of the radio frequency spectrum for operating the cellular system. AT&T explained to the FCC that if a 75-MHz spectrum band could be allocated to the cellular system, the system would provide unlimited capacity for voice to an unlimited number of users.

In 1974, the FCC allocated 40 MHz to the cellular system and reserved 30 MHz for it. The FCC allocated less spectrum to the cellular system because during its inception it could only handle a few subscribers and needed less spectrum.

In 1973, Frank Brecher at Bell Labs led the cellular commercial system development team developing the radio equipment. Naperville Bell Labs in Chicago completed the mobile switch equipment by modifying No. 1 electronic switching system (ESS) with No. 1A processor. The modified No. 1 ESS for a cellular switch was finished in only 8 months. A decisive change, besides the calling process and handoff functions, was the local line number (LLN). In mobile switching, there are no fixed LLNs associated with any frequency channel. Users can use different frequency channels on different calls. Therefore, a pseudo LLN was used to replace the LLN. This decisive change saved the existing software from a major change or modification. The following key technologies are unique to the cellular systems.

1. The difference in mobile phone design from the wire-line phone overcomes the signal fading received in the mobile radio environment. The signal goes up and down as the vehicle moves. As the vehicle travels faster, the signal strength changes more rapidly. The rapid change in signal strength degrades voice quality. A "diversity scheme" is used to reduce fading by receiving two different fading signals, both of which carry the same voice information and combining them to smooth out fading. The most popular among the diversity schemes is space diversity, which receives two fading signals from two separate antennas. At the mobile station, two antennas are separated by a half wavelength or greater[9] (i.e., roughly 6 in at 850 MHz and 3 in at 1900 MHz). At the base station, two antennas are separated by eight wavelengths[10,11] (i.e., roughly 9 ft at 850 MHz and 4 ft at 1900 MHz). Now after 15 years of cellular operation we have proven that the diversity scheme is a key element in improving the system performance for both voice and data in the cellular system.

2. *Frequency reuse scheme.* To increase the spectrum efficiency, the same frequency channel can be reused in different locations. Based on the voice quality requirement (i.e., to reduce the cochannel interference), the minimum separation in space for two cochannels (same frequency) is specified by a parameter q[3,12]

$$q = \frac{D}{R} \qquad [1.1]$$

where D is the cochannel-cell separation and R is the cell radius. In advanced mobile phone services (AMPS), q is 4.6. This means that all the adjacent cochannel cell sites have to be $D = 4.6R$ apart. From the hexigon topology configuration shown in Fig. 1.1, the cluster of seven cells is calculated by the frequency reuse factor K[3,12]

$$K = \frac{1}{3} \cdot \frac{D}{R} = \frac{1}{3}q^2 = \frac{1}{3}\left(\frac{D}{R}\right)^2 = \frac{1}{3}(4.6)^2 = 7 \qquad [1.2]$$

In a cluster of seven cells, every cell is assigned a different frequency. If a system can make q a smaller value than 4.6, while maintaining voice quality, from Eq. [1.1], K becomes less than 7. The smaller q is, the higher the radio capacity, and the smaller K is.

For a high traffic area, each call needs to further divide into three sectors, and each sector has a different set of frequencies, as shown in Fig. 1.1. The frequency reuse scheme

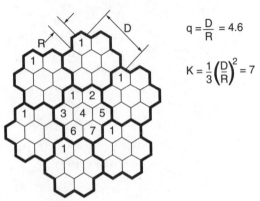

$$q = \frac{D}{R} = 4.6$$

$$K = \frac{1}{3}\left(\frac{D}{R}\right)^2 = 7$$

Figure 1.1.   Frequency reuse pattern.

is the key to increasing the capacity for the cellular system.[13] Because of frequency reuse, the adjacent cells operate on a different set of frequencies. Each frequency channel that carries a call in one cell will be handed off to another frequency channel in another cell while the mobile station is traveling to the new cell during the call. This process is called a handoff. The system engineers designed a handoff process controlled by the system so the user would not have to intervene. This was a new, unproven technology in the early 1970s. Its development involved a high risk and a high cost. However, the success of cellular business was dependent on the handoff feature, which allows the customer to talk on the mobile phone while driving, no matter how far, without experiencing a cell drop.

1.4 AT&T'S SUCCESSFUL MARKETING STRATEGY IN 1963 FOR SATELLITE COMMUNICATION

In 1960 the Echo project, which was the first satellite communication experiment using a balloon-type satellite (i.e., a passive satellite) was successfully completed. It was followed by John Pierce's theoretical study. The voice communication link through a passive satellite between Crawford Hill Bell Labs and JPL California performed very well. Then the Telestar (i.e., an active satellite) was successfully launched. The satellite communication business would be monopolized by AT&T in the future. AT&T conducted a great marketing strategy in promoting satellite communications between 1961 and 1964. AT&T's marketing created such a big wave of interest in the United States before the commercial deployment that it caused Congress to realize the importance and impact of this business. Congress stopped AT&T from engaging in the satellite communication business. Instead, a new company known as Comsat was formed. AT&T was ordered to give all its technology and financial support to Comsat. In the interest of the nation, this was a good move. However, it was not beneficial to AT&T's

marketing strategy, which should have been changed to delay its advertisement until the satellite communication business was underway and then gradually promote the business. With this strategy, AT&T would have enabled the satellite communication business to grow steadily before the government learned about it and started getting involved.

Satellite communication was now a global business. Its technologies were dominant in the 1970s, but satellite communication does not involve personal communications directly, and the markets for personal communications are huge. It is why the cellular communications/personal communications service (PCS) started to take off in 1980 and Internet communications started to take off in 1990. Nevertheless, we have to give the credit to AT&T's early contribution in satellite communication.

1.5 WHY THE CELLULAR SYSTEM COULD NOT BE COMMERCIALLY DEPLOYED IN THE 1970S

While the development of AMPS was being completed by AT&T Bell Labs in 1976, the FCC allocated 40 MHz of spectrum for cellular use. However, the FCC couldn't release the cellular license to AT&T because the Radio Common Carrier (RCC), an association of dispatch and paging, enacted a strong lobby against AT&T to block them from running a cellular business. The RCC reasoned that the cellular business would directly threaten their business. Their fears were realized once the cellular business took off, but they managed to keep the FCC from making any moves until 1980. At that time, the FCC decided to split the 40-MHz spectrum band into Band A and Band B. Each of these bands was 20 MHz. Band A was licensed to non-wireline (dispatch and paging) companies, and Band B was licensed to the wire-line (telephone) companies. After this settlement, the cellular systems were deployed in 1983. This is why even though the United States was the first to invent the cellular system, the Japanese NTT version of AMPS was the first cellular system in the world, deployed in Tokyo in 1979.

RCC's visions were right in the late 1970s; as we can see, the cellular industry is booming today. The market penetration has reached 15 percent and still keeps growing. However, the Band A companies should not have sold their business so early. Today most Band A companies have been bought by the Band B companies. A history of these events is outlined in Sec. 5.5.

1.6   WHY OKI GOT THE FIRST 200 CELLULAR PHONE DEAL

AT&T had three functions, operational (21 regional operators), manufacturing (Western Electric Co.), and research (Bell Laboratories). The U.S. Department of Justice (DOJ) had planned for a long time to divest AT&T. But AT&T was always negotiating to keep the three functions, operation, manufacture, and research, enacted. In 1974 DOJ banned AT&T from manufacturing cellular phones. Then an open bid to the outside manufacturers from AT&T was announced. Among RCA, Motorola, E.F. Johnson, GE, and others, a Japanese company, Oki, was interested in bidding. After World War II, the U.S. government gave Japanese companies a special provision that allowed them to enjoy an equal opportunity in manufacturing bidding in the United States. Oki won U.S. Congress support to obtain the business. The first 200 handsets Oki made used AT&T's technology, which Oki learned. Then 6 months later another bid for an additional 1800 handsets was announced by AT&T. This time three companies won the bids: E.F. Johnson, Oki, and Motorola. Each of them manufactured 600 units.

To test the phones made by the outsources, Bell Labs developed test equipment to ensure that all the specified equipment parameters for the handset were met. Creating this test-equipment concept was very important because it made sure that one manufacturer's base station could operate on another manufacturer's handset. All parameters needed to fall in their specified ranges. Later, the test-equipment concept was applied in Global System Mobile, or GSM (originally named Special Mobile Group), commercial equipment development.

1.7   THE FAST-FADING MODEL AND FIELD-COMPONENT DIVERSITY

In 1965, John Pierce, executive director of Bell Labs, asked E. N. Gilbert, the famous coding and communication theory expert, to study his new idea[14] of combining the E H fields to reduce fast fading. Gilbert created the multipath fading model for analyzing the performance of the so-called energy density signal, denoted as Wm:

$$Wm = \varepsilon|E|^2 + \mu|H|^2 \qquad [1.3]$$

where ε is the permittivity of medium, or called dielective constant, and μ is the permeability of medium. Gilbert's paper was published in 1965.[15] W. C. Y. Lee took Gilbert's model and derived the second-order statistics, such as level crossing rate, duration of fades, and the power spectrum published in 1966.[16] Jakes and Ruddink applied Lee's derivation in their paper in 1966.[17] In 1965, R. H. Clarke called meetings on behalf of Cutler to ask for formal presentations on theoretical, statistical, and experimental approaches from three researchers (see Exhibit 1.A). In 1967, R. H. Clarke[18] wrote an outstanding summarized paper using Gilbert's model. Most of the second statistics in Clarke's paper came from Lee's. Today, many authors in this field do not know its history. Correctly, we should name today's multipath mobile model the Gilbert model to honor his pioneer contribution. Of course this field-component diversity idea of using the energy density concept needed an energy density antenna, which was Lee's design.[19] The antenna is a cross semicircle loop antenna, as shown in Fig. 1.2. A 180° phase difference hybrid is used. The sum port collects the E field and the substrating port collects the H field. In 850 MHz, the diameter of the loop is 1.5 in. This antenna was used to prove the concept. The measurement showed that the E and H fields are uncorrelated when they are simultaneously received. This was a very good diversity scheme because no antenna separation was required. The only small drawback was that the loop gain was weaker than the dipole antenna. Also, the requirement for the antenna spacing of

BELL TELEPHONE LABORATORIES
INCORPORATED

SUBJECT: **Discussions on Some Aspects of Mobile Radio**	DATE: **May 12, 1965**
	FROM: **R. H. Clarke**

MESSRS. C. C. CUTLER:
     E. N. GILBERT:
     W. C. JAKES:
     W. C. Y. LEE: ←| THIS COPY FOR |

It has been proposed that a few of us who are directly concerned with the interpretation of mobile-radio experiments and the analysis of possible mobile-radio systems should get together, at least occasionally, for the discussion and criticism of any ideas that may be current. It is envisaged that a useful format for such sessions would be to have half the available time devoted to a formal presentation of a topic or group of topics, with the rest of the time taken up with discussion. It should rapidly emerge whether this is the best format or not, and what the frequency of these meetings should be.

The first meeting has been scheduled for 9 a.m. on Wednesday, May 19, at Murray Hill in Mr. C. C. Cutler's office, room 1E-337. On this occasion the undersigned will give the formal presentation.

This presentation will begin with a brief review of some results obtained over the past few years with experimental mobile-radio systems in the vicinity of Murray Hill, together with an outline of some new systems which are in the process of being tested. The basis of a simple theoretical approach, which takes as its starting point a statistical description of the radio-frequency fields incident on the mobile receiver, will be explained. This approach will then be used to calculate the "fading" spectrum of the detector output of the receiver. Finally some experimental fading spectra will be compared with these theoretical results, and a few predictions will be made of the fading spectra likely to be encountered in the systems now under tests.

ORIGINAL SIGNED BY

HOH-1351-RHC-EEB   R. H. CLARKE

Exhibit 1.A.   R.H. Clarke's memorandom.

---

space diversity on the vehicle roof only needs one-half of a wavelength (6 in at 800 MHz), as stated in Sec. 1.3. It was easy to implement. Thus the energy density antenna for the diversity scheme was put on hold. In the late 1980s, the mobile handsets were produced and the diversity schemes were needed. Due to the small amount of space at the handset, the E-H field

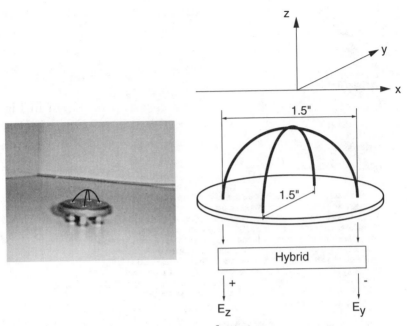

Figure 1.2.   Energy density antenna configuration.

diversity scheme concept started to come back and was applied to the mobile handsets in Japan.

1.8   THE FIRST PROTOTYPE MOBILE UNIT AND CELL SITE

The prototype mobile unit was developed in 1970. It was different from the Improved Mobile Telephone Service (IMTS) system in the 1970s. First, it needed to generate any one of several hundred RF channels at the mobile unit upon command from the land network. A sophisticated frequency synthesizer was required. Second, the mobile unit used diversity to protect the operated RF channel against Rayleigh fading. Third, the high-level integration of new circuit technologies was employed to reduce the mobile phone cost and size.

A prototype cellular mobile telephone unit was created that was about $12 \times 12 \times 25$ in. The unit consisted of a collection of epoxy–fiberglass printed circuit boards mounted in the cavities of an aluminum casting. The casting provided not only a rugged, low-cost enclosure, but also sufficient shielding to prevent the generation of spurious signals, a problem often encountered with frequency synthesizers.

The mobile unit, a block diagram of which is shown in Fig. 1.3, was a sophisticated frequency modulation (FM) transceiver that provided duplex voice transmission and reception by division of the RF band into two segments separated by the IF frequency so that one frequency-generating system may serve for the source of both transmitter and local oscillator power. The final project was led by Reed Fisher at Bell Labs.[20] Figure 1.4 shows a cellular analog handset phone called Startec, made by Motorola

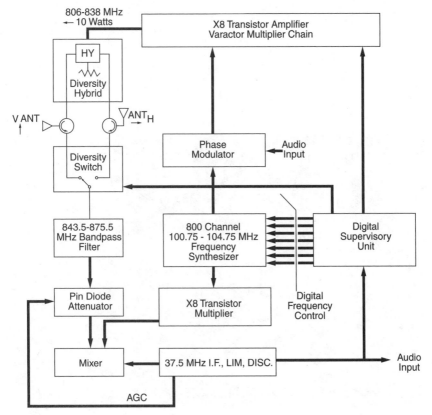

Figure 1.3. A block diagram of high-capacity mobile unit.

Figure 1.4. Motorola's Startec, a cellular analog handset.

Figure 1.5. Interior view of Lyons cell site building.

Figure 1.6. A Hyundai pico cell in 1999.

in 1995. The size of this phone is drastically smaller than the previous prototype and has many more features.

Figure 1.5 shows the cell site of a 16-channel radio equipment set up at the Lyons cell site building at Lyons, Illinois, in 1977. The 16-channel combiner is shown to the left in the figure. The size of the cell site was very large as compared with the pico cell made by Hyundai in 1999, shown in Fig. 1.6

1.9 THE PIONEER SYSTEM DEPLOYMENT TOOL FOR REGIONAL BELL OPERATING COMPANY (RBOC) START-UP MARKETS

In 1984 the cellular system deployment tool provided by AT&T at Bell Labs, Whippany, N.J., was based on Lee's propagation prediction model. The verification of the new propagation model was started in 1974. The model used the terrain maps from the Defense Map Agency (DMA) to generate the prediction signal reception data points. For better resolution, 50,000:1 scale topographic maps were used. Each map size was roughly 5 by 8 miles. The contour lines were in 20-ft increments. To generate a grid map and store the elevation of each in the computer, each grid was about a half a square kilometer (18 × 24 grids) on the map. Then an eyeball average of the terrain altitudes in each grid was used. The eyeball average values turned out to be good enough to use for Lee's model predictors. This project was carried out in a joint effort with Bell Labs and the TriState Team (New York Bell, New Jersey Bell, and New England Bell). The averaging elevation process for the grid map was labor intensive.

Later, the resolution of the contour map was relaxed. Then the alternative was to purchase the Defense Map Agency Topographic Center (DMATC) 1 × 1° tape from DMA.[21] It had 120 × 120 data grids. The size of each data grid was 3 × 3″. As shown in Fig. 1.7, the shape of 3 × 3″ grid at the earth's equator is a square. However, at a higher latitude away from the equator, this grid is not a square. For example, at a latitude of 40°, the size of the grid (longitude-latitude) is 200′ × 300′. The altitude value in each grid was stored in the tape from a 250,000:1 digitized map. The topological contour resolution of such a map is 200-ft increment (i.e., 0-, 200-, 400-ft contours, and so forth). The altitude in each grid was obtained by extrapolating from the coast topological contours and then taking an average of 30 altitude values from 30 grids (5 × 6 grids). Use this value to compare with the value obtained from the eyeball average at the same location, the result was that the accuracy obtained from two methods was the same. Because the eyeball average was labor intensive, the averaging altitudes from DMA 1 × 1° tape were accepted. As the computer memory space became large and commercially the whole 1200 × 1200 grid could be stored in the computer and used for input to Lee's propagation prediction model. By March 1979, the FCC was still not able to issue the cellular system license to AT&T,

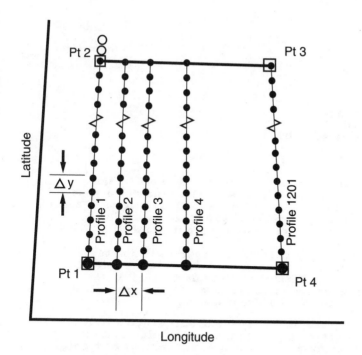

$\triangle x$ = 3 arc-seconds
$\triangle y$ = 3 arc-seconds
O = Elevation point in adjacent 1° block
● = Elevation point
⬤ = First point along profile
☐ = Corner of DEM polygon
    (1° block, scale 1:250,000)

Figure 1.7. Structure of a 1:250,000 scale digital elevation model.

although the AMPS system design was finished. Lee left Bell Labs and joined the ITT Defense Communication Division.

Lee wrapped up the "Lee model" work and handed it to Bell Labs before leaving. On October 30, 1979, Bell Lab patent attorney C. S. Phelan wrote a letter to Lee to acknowledge receipt of Lee's memorandum: "A New Mobile Radio Propagation Model Case 39445-7," dated March 30, 1979. The letter also acknowledged the appreciation of R. D. Johnson, with whom Lee worked at Bell Laboratories Mobile Communication Labs (see Exhibit 1.B).[22]

Bell Laboratories

600 Mountain Avenue
Murray Hill, New Jersey 07974
Phone (201) 582-3000

OCT   30 1979

Dr. W. C. Y. Lee
9 Hickory Road
Denville, New Jersey   07834

Dear Bill:

In reply to your letters of October 17 and 21, 1979, this
is to advise you that we have now received the ribbon copy
of your memorandum "A New Mobile Radio Propagation Model
Case 39445-7" dated March 30, 1979. That copy includes
photocopies of the 29 figures mentioned in the text.

It is my pleasure to relay to you the appreciation of the
technical personnel with whom you worked in the Bell
Laboratories Mobile Communications Laboratory. In the
course of recent communications regarding the propagation
model, Mr. R. D. Johnson wrote to me:

> "Bill Lee, in his 15-year career with Bell
> Laboratories, has made many significant
> contributions toward advancement of mobile
> radio technology. His numerous publica-
> tions attest to his productive career at
> Bell Laboratories."

> "If you should have additional opportunities
> to communicate with Bill, please indicate
> that we very much appreciate his contribu-
> tions to our work in Mobile Communications.
> We are particularly grateful for his spear-
> heading work in the development of the new
> mobile propagation model for use in planning
> the Bell System's new Advanced Mobile Phone
> Service. Please give Bill our best regards
> and best wishes for continued success."

Finally, Bill, please accept my personal best wishes that
you find both success and fulfillment in your new employment.

Very truly yours,

C. S. Phelan
Patent Attorney

CSP:jc

Exhibit 1.B.   C. S. Phelan's letter to Lee.

1.10   DIGITAL CELLULAR SYSTEMS

1.10.1   TDMA Systems

Because the cellular system started to grow very rapidly, the limitations of analog system capacity were of concern. One way to increase the capacity is to go digital. The first-generation system is analog, or we may say the frequency division multiple access (FDMA) system. The second-generation system is the digital time division multiple access (TDMA) system. The third-generation system is the digital code division multiple access (CDMA) system.

1.10.2   NATDMA and PDC[23–28]

The North American TDMA (NA-TDMA) and Personal Digital Cellular (PDC) use TDMA technology to achieve a high-capacity system. Specification IS-54 was NA-TDMA, and later modified and renamed IS-136. NA-TDMA systems were mainly deployed by AT&T, SBC, and Bell South in the United States in 1993 and PDC was deployed in Japan in 1994. Although these two systems were successfully deployed, neither can demonstrate superiority over other digital systems in radio capacity. They are about three times over the capacity of AMPS.

1.10.3   GSM[29–31]

The first digital cellular system started in development in 1983 by a group called GSM (special mobile group), sponsored by the European Community. The group was working on a digital system, which was different from the AMPS, an analog system. Also, they wanted to challenge the advanced technology using TDMA multiple access. TDMA, used in GSM, was the first technology applied to a cellular (mobile radio) system. At that time, the group did not consider the radio capacity issue. Nobody could imagine in the early 1980s that the radio capacity could be a big issue later due to fast growth in the cellular industry. The network of GSM was used by modifying from the wire-line Advanced Intelligent Network

(AIN). The wire-line AIN only dealt with a fix-to-fix link. When one end was moving instead of fixed, the network had to act fast to establish the call and maintain it while the mobile unit was traveling; otherwise, the call would be dropped. After 35 revisions, the GSM network became the most intelligent network in cellular operations.

The successful Intelligent Network (IN) made GSM a leading system in worldwide penetrations. In Europe, the development of the wireless IN needed a collective effort. The engineers from all European countries worked together, contributing their individual wisdoms to a big project. In the United States, innovation is on everyone's mind and trying to achieve it to be a hero is paramount. Only the Old Ma Bell system put a collective effort into the United States' nationwide network. This raises a question of whether regulating or divesting AT&T is right. If the answer is the latter, the price we pay is the missing collective effort in building big projects in the United States in the private sector.

1.10.4   CDMA SYSTEMS: CDMAONE[32–35]

The CDMA system was developed by Qualcomm Co. in 1989 after NA-TDMA was voted a second-generation standard. The CDMA system was also treated as a second generation and widely deployed in the United States, Korea, and Hong Kong. It was demonstrated as a high-capacity system. The Qualcomm-developed CDMA system was renamed cdmaOne by the Cellular Development Group (CDG) in 1997. A detailed description of CDMA appears in Chap. 6.

1.11   LOW-ORBIT SATELLITE MOBILE[36]

In the past, geostationary satellite systems have been deployed. It takes only three or four geostationary satellites to cover the entire earth. These satellites rotate around the earth at the same speed as the earth rotates. Thus, the earth-to-satellite link is always fixed, and it is easier for satellite communications. However, with satellite-mobile communications, the power of mobile transmission is limited. The transmit power of a mobile handset is further limited. Also, the 250-ms round-trip time for voice response is too long. The system latency issue is always a concern. Therefore the low-earth-orbit satellite (LEO) was introduced first in the 1960s at Bell Labs. It was called the Low Attitude Active Communication Satellite[37] and was proposed to the National Aeronautics and Space Administration. The satellite system carried a two-way wideband signal at an altitude between 600 and 5000 nautical miles and had an elliptical orbit.

The advantage of implementing LEO is to use a low-power device and have low latency. Forty to 70 satellites would carry more capacity by frequency reuse and cover the entire world. The Bell Labs' proposed LEO was never deployed.

In late 1980, LEO started to come back for the satellite-mobile system. Two popular systems are Iridium and Globalstar. The Iridium system[38] consists of 66 satellites, each of which acts as a base station as well as a switch in space. The Iridium system is an advanced system that uses military technology. Mobile initiation calls can be received by the Iridium satellites, which rotate around the earth about every 2 hours. The Iridium satellites switch the call from satellite to satellite until they reach the destination and then send it down to the ground gateway. Iridium needs fewer gateways than the Globalstar does. However, the cost of this system is very high, and some countries are concerned about their sovereignty as well as losing control of their national communication network.

The Globalstar system[39] consists of 48 satellites, each of which is a relay. It may also be called beacon or repeater. The altitude of Globalstar satellites is higher than that of the Iridium satellites. Thus it requires fewer satellites to cover the entire world than does the Iridium system. The mobile initiating calls are received by the Globalstar satellites. Because the satellites are relays, the calls are sent down to the gateway as soon as they receive the calls. The gateways connect to the land-line network of each country. Because relay operation is in the satellites, the required number of gateways for Globalstar is higher than for Iridium. However with the Globalstar system the sovereignty issue has disappeared due to the nonswitching function in the satellites. The network is also simple, and its cost is low.[37] LEO is further discussed in Sec. 5.21.

1.12   REFERENCES

1. Bell Laboratories, "High-Capacity Mobile Telephone System Technical Report." Submitted to FCC December 1971.
2. Bell Laboratories, "High Capacity Mobile Telecommunications System" Developmental System Reports Nos. 1–8 published every 3 months, from March 1977 to March 1979.
3. V. H. MacDonald, "The Cellular Concept," *Bell System Technical Journal*, vol. 58.
4. F. H. Blecher, "Advanced Mobile Phone Services." *IEEE Trans. on Vehicular Technology*, vol. VT-29, May 1980, pp. 238–244. Jan. 1979, pp 15–42.
5. W. C. Y. Lee, *Mobile Cellular Telecommunications: Analog and Digital Systems*, 2d ed., New York: McGraw-Hill, 1995.
6. W. D. Lewis, "Coordinated Broadband Mobile Phone Systems." *IEEE Trans. Veh. Tech.* Comm. VC-9, May 1960, pp. 43–48.
7. D. O. Reudink, "Comparison of Radio Transmission at X-Band Frequencies in Suburban and Urban Areas," *IEEE Trans. Ant. Prop.* AP-20. July 1972, p. 470.
8. C. L. Ruthroff and L. U. Kible, "A 60 GHz Cellular System." *Microwave Mobile Symposium*, Boulder, Colorado, 1974.
9. W. C. Y. Lee, "An Extended Correlation Function of Two Random Variables Applied to Mobile Radio Transmission," *Bell System Technical Journal*, vol. 48, Dec. 1969, pp. 3423–3440.
10. W. C. Y. Lee, "Antenna Spacing Requirement for a Mobile Radio Base-Station Diversity," *Bell System Technical Journal*, vol. 50, July–August 1971, pp. 1859–1874.
11. W. C. Y. Lee, "Effects on Correlation Between Two Mobile Radio Base-Station Antennas," *IEEE Trans. Comm.*, vol 21, Nov. 1973, pp. 1214–1224.
12. W. C. Y. Lee, *Mobile Communication Design Fundamentals*, 2d ed., John Wiley, 1993.
13. R. H. Frenkiel, "A High-Capacity Mobile Radio Telephone System Model Using a Coordinated Small-Zone Approach," *IEEE Trans. Veh.* VT-19, May 1970, pp. 173–177.
14. J. R. Pierce, "Fading in Mobile Radio–Case 22108-11," Bell Lab internal memorandum for record, October 22, 1964.

15. E. N. Gilbert, "Energy Reception for Mobile Radio," *Bell System Technical Journal*, vol. 44, October 1965, pp. 1779–1803.

16. W. C. Y. Lee, "Statistical Analysis of the Level Crossings and Duration of Fades of the Signal from an Energy Density Mobile Radio Antenna," *Bell System Technical Journal*, vol. 46, February 1967, pp. 417–448.

17. W. C. Jakes, Jr., and D. O. Reudink, "Comparison of Mobile Radio Transmission at UHF and X-Band," *IEEE Trans. Veh. Tech. 16*, October 1967, pp. 10–14.

18. R. H. Clarke, "A Statistical Theory of Mobile Radio Reception," *Bell System Tech. Journal*, 47, July 1968, pp. 957–1000.

19. W. C. Y. Lee, "An Energy Density Antenna for Independent Measurement of the Electric and Magnetic Field," *Bell System Technical Journal*, vol. 46, Sept. 1967, pp. 1587–1599.

20. Reed Fisher, "A Subscriber Set for the Equipment Test," *Bell System Technical Journal*, vol. 58, Jan. 1979, pp. 123–144.

21. U.S. Department of the Interior, Geological Survey, "Digital Terrain Tapes," National Cartographic Information Center, US Geological Survey, 507 National Center, Reston, Virginia.

22. C. S. Phelan, Bell Lab Patent Attorney. A letter to W. C. Y. Lee to appreciate "A New Mobile Radio Propagation Model Case 39445-7," March 30, 1979.

23. Cellular System, IS-54 (incorporating EIA/TIA 553), "Dual-Mode Mobile Station-Base Station Compatibility Standard," Electronic Industries Association Engineering Department, PN-2215, December 1989 (NADCA-TDMA system).

24. Cellular System, IS-55, "Recommended Minimum Performance Standards for Mobile Stations," PN-2216, EIA, Engineering Department, December 1989 (NADC-TDMA system).

25. Cellular System, Minimum Performance Standards for Base Stations," PN-2217, EIA, Engineering Department, December 1989 (NATC-TDMA system).

26. Cellular system, IS-136 "800 MHz TDMA Cellular-Radio Interface-Mobile Station-Base Station Compatibility, (1) Digital Control Channel (2) Traffic Channels and FSK Control Channel." TIA/EIA, Dec. 1994.

27. Cellular Systems TIA/EIA/IS-137 "800 MHz TDMA Cellular-Radio Interference-Minimum Performance Standards for Mobile Standards," TIA/EIA, Dec. 1994.

28. Cellular System TIA/EIA/IS-138 "800 MHz TDMA Cellular-Radio Interface-Minimum Performance Standards for Base Station," TIA/EIA, Dec. 1994.

29. "European Digital Cellular Telecommunications System (Phase 2): General Description of a GSM Public Land Mobile Network," ETSI, 06921 Sophia Antipolis Cedex, France, October 1993, GSM 01–12.

30. *Proc. Third Nordic Seminar on Digital Land Mobile Radio Communication*, September 12–15, 1988, Copenhagen (21 papers describe the GSM system).

31. Bernard J. T. Mallinder, "An Overview of the GSM System, *Proc. Digital Cellular Radio Conference*, Hagen FRG, October 1988.

32. Cellular System, IS-95, "Dual-Mode Mobile Station-Base Station Wideband Spread Spectrum Compatibility Standard," PN 3118, EIA, Engineering Department, December 1992 (CDMA system)

33. Cellular System, IS-96, "Recommended Minimum Performance Standards for Mobile Stations Supporting Dual-Mode Wideband Spread Spectrum Cellular Base Stations," PN-3119, EIA, Engineering Department, December 1993 (CDMA system).

34. Cellular System, IS-97, "Recommended Minimum Performance Standards for Base Stations Supporting Dual-Mobile Wideband Spread Spectrum Cellular Mobile Stations," PN-3120, EIA, Engineering Department, December 1993 (CDMA system).

35. W. C. Y. Lee, "Overview of Cellular CDMA," *IEEE Trans. on Vehicular Technology*, May 1991, p. 291–302.

36. W. C. Y. Lee, *Mobile Communications Engineering, Theory and Applications*, 2d ed., New York: McGraw-Hill, 1998. pp. 540–547.

37. Bell Labs proposal to National Aeronautics and Space Administration. "Low Altitude Active Communication Satellite" for Proposal No. GS-1861, March 20, 1961.

38. J. E. Hatlelid and L. Casey, "The Iridium System: Personal Communications Any-Time, Any-Place," *Proc. Third International Mobile Satellite Conference*, Pasadena, June 16–18, 1993, pp. 285–290.

39. R. A. Wiedeman, "The Globalstar Mobile Satellite System for Worldwide Personal Communication," *Proc. Third International Mobile Satellite Conference*, Pasadena, June 16–18, 1993, pp. 291–296.

1.13   READING MATERIAL

Dixon, R. C., *Spread Spectrum Systems*, 3d ed., New York: John Wiley, 1994.

Feher, Kamino, *Wireless Digital Communications*, Prentice-Hall, 1995.

Gallagher, M., and R. Snyder, *Mobile Telecommunications Network*, New York: McGraw-Hill, 1997.

Jakes, W. C., ed., *Microwave Mobile Communications*, New York: John Wiley, 1974.

Lee, W. C. Y., *Mobile Cellular Telecommunications, Analoged Digital Systems*, 2d ed., New York: McGraw-Hill, 1995.

Lee, W. C. Y., *Mobile Communications Design Fundamentals*, 2d ed., New York: John Wiley, 1993.

Lee, W. C. Y., *Mobile Communications Engineering, Theory and Applications*, 2d ed., New York: McGraw-Hill, 1998.

Mouly, M., and M. B. Pautet, *The GSM System for Mobile Communications*, M. Mouly & M. B. Pautet, 1992.

Ojanpera, T., and R. Prasad, eds., *Wideband CDMA for Third Generation Mobile Communications*, Artech House, 1998.

Rappaport, T. S., *Wireless Communication*, Prentice-Hall, 1996.

Simon, M. K., J. K. Omura, R. A. Scholtz, and B. K. Levitt, *Spread Spectrum Communications Handbook*, New York: McGraw-Hill, 1994.

Smith, C., and C. Gervelis, *Cellular System*, New York: McGraw-Hill, 1996.

Stüber, G. L., *Principles of Mobile Communication*, Kluwer Academic Publishers, 1996

Viterbi, A. J., *Principles of Spread Spectrum Communication*, Addison-Wesley, 1995.

WHY MOBILE RADIO SYSTEMS ARE DIFFICULT TO DEVELOP

2.1   LIMITED NATURAL SPECTRUM

The spectrum of electromagnetic waves is a limited natural resource, so using the spectrum efficiently is a big challenge. In wireless communications, radio interference within its own allocated spectrum and in neighboring spectrums limits the number of operational service systems.

In the 1970s we had hoped a new discovery, a gravitational wave, might provide a new domain of spectrum and help us to open another means of communication. The gravitational wave could supposedly be propagated through the metal plate. But, the experimental data could not be reproduced under the same condition. Our hopes dimmed. Of course, we still do not believe that acoustic and electromagnetic are the only two waves that exist in the universe that can be used for wireless communications. We may find a third type of wave in the future that could have more spectrum and travel faster than the speed of light. But for today, we must continue to manage our limited spectrum resources.

As Internet services became worldwide businesses, the wireless Internet services also began to take off. For the wire-line Internet services, high-speed data (up to Gbps) requiring a 10-GHz spectrum bandwidth can be easily carried by a fiber link. However, the spectrum bandwidth is a very precious commodity for wireless Internet services because the limited natural spectrum (even 2-Mbps rate is hard to provide). Therefore we have to use wireless for high-speed data service only for portability or nomadic application. Yet we still depend on fiber for large bandwidth transmission. The outcome is a hybrid network mixed with both wire-line and wireless transmission. The wireless section can be the last mile or last 100 m or last 50-m link, depending on the desired data rates. Given a spectral bandwidth, the higher the data rate, the shorter the link range.

2.2   WHY WE NEED A CARRIER

In radio communication, we have to transmit a signal through the air. To transmit a signal, an antenna the size of at least half a wavelength is needed. If the information signal (voice or data) at the baseband is at a low frequency (in kilohertz or megahertz), it will be very difficult to transmit through the air.

At a low frequency, its wavelength is too large, and the half-wavelength antenna is impractical to construct. For example, for a frequency of 10 kHz the antenna size should be 15,000 m (1/2 wavelength), which is impossible. Instead we send information by a higher frequency, which we call the carrier (i.e., carry the information signal through the air with a smaller antenna). Usually, at the base station the size of an antenna can be larger (up to 50 ft). With mobile, the antenna size should be small (less than 6 in). The size of the antenna is referred to as the wavelength of the carrier frequency. If the antenna size is greater than 1/2 wavelength, there is a power gain generated over the gain of a half-wavelength antenna. In a 800-MHz carrier, the half wavelength is 6 in. This antenna size can be handled easily by a handset, but we cannot let a carrier frequency get too high, because a higher carrier frequency causes a greater propagation loss. Over a 10-GHz frequency, additional atmospheric (rainfall rate) loss occurs. At 20 GHz there is a drastic loss due to the absorption of water vapor particles. Another drastic loss occurs at 60 GHz due to the absorption of oxygen molecules, as shown in Fig. 2.1.[1] Therefore, in the mobile radio system, the carrier frequency should stay below 10 GHz. Much research was carried out in the 1970s to demonstrate that mobile communications could be carried out at or below 10 GHz.

2.3   WHAT IS THE MOBILE RADIO ENVIRONMENT?[2]

Because the mobile antenna height of a mobile station is very close to the earth's ground, the following affect the mobile signal (receiving and transmitting):

1. The natural variation of terrain configuration
2. The human-made effect, including the human-made structures, such as urban, suburban, high-rise buildings, and bridges, and human-made noise, such as automobile ignition noise, and industrial noise

Pressure: 1 atm
Temperature: 20° C
Water Vapor: 2.5 g/m^3

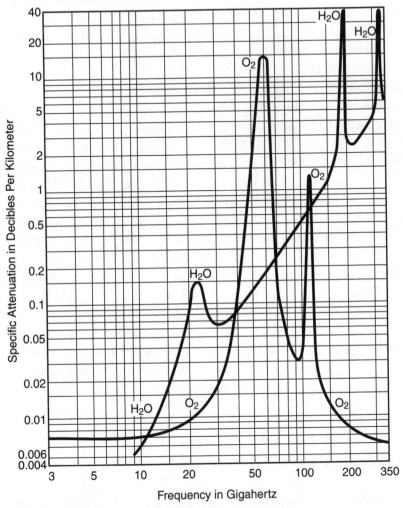

Figure 2.1. Specific attenuation due to atmospheric gases.

2.3.1 NATURAL TERRAIN VARIATION

As a mobile station moves, the terrain makes the mobile signal strong or weak depending on its configuration (i.e., a high or low spot). The low-height mobile antenna also receives two waves from the base station antenna: a direct wave and a ground-reflected wave. These two waves added together cause a d^{-4} propagation loss which is greater than a d^{-2} propagation loss (a free-space or high tower propagation loss) over a distance.

2.3.2 Human-Made Structure

The mobile station is situated in a human-made structure that causes the signal to be reflected back and forth before reaching the mobile unit. These multipath waves introduce the signal fading phenomenon as the mobile moves. It reduces the average carrier power of the receiving signal. It also degrades the voice quality and data performance.

2.3.3 Time Delay Spread

The multipath waves also cause an echo phenomenon. These waves arrive at the mobile unit at different times, normally within 100 μs. The echo phenomenon is called the *time delay spread* in radio communications.[3] In an analog voice signal (3000 Hz) 1 Hz travels 0.33 ms, so the short-arrival multipath echo phenomenon won't be noticed over a long time interval voice cycle. In high-speed data, over 100 Kbps, the bit interval is 10 μs. The echo phenomenon based on the arrival of multipath waves can be noticeable and interfering. We measure the time delay spread[3] from the time delay of the first wave to the last reflected wave arrival, as shown in Fig. 2.2. The easiest way to measure the delay spread is to take 30 dB below the first wave arrival. Usually the first wave arrival is at a short distance, as shown in Fig. 2.2*a*; thus less propagation loss occurs to the wave. However, due to the absorption of the reflection surface, the first wave arrival may not be the strongest, as shown in Fig. 2.2*b* and *c*.

The time delay spread limits the data transmit rate. To transmit high-speed data, we have to cancel the multipath reflected waves by applying an equalizer. Most time-using diversity schemes can also reduce the time delay spread interval (see Sec. 7.14).

2.3.4 Human-Made Noise

The human-made noise within the mobile-radio environment is primarily from unintentional sources, such as vehicular

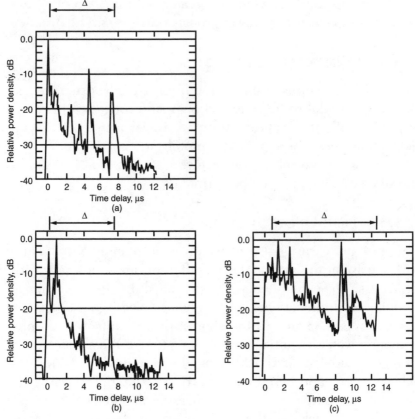

Figure 2.2. Delay envelopes in New York City. (*Reproduced with permission from I.E.E.E.*)

ignition, radiated noise from power lines, and industrial equipment. The human-made noise has been classified into three categories:

1. Urban areas
2. Suburban areas
3. Open areas

The noise floors in different cities and in different frequency bands are different. The noise floor in an urban area can be 15 dB higher than that in a suburban area. In general, the higher the frequency, the lower the noise floor.

The major source of the human-made noise is vehicular ignition noise. An eight-cylinder engine with a 3000-rpm speed can generate 200 spikes per second. Each spike in the time domain spreads its energy across a wideband in the frequency domain. This is just from one vehicle. The noise floor from a traffic density of 1000 vehicles per hour would be higher than that of 100 vehicles per hour[4] (as shown in Fig 2.3). It explains why the noise floor in urban areas is higher than in suburban areas. Also, the human-made voice floor is lower as the frequency is higher. At a frequency of 1 GHz, the human-made noise floor of 1000 vehicles per hour is 5 dB above the thermal noise, but at 100 MHz, the noise floor of 100 vehicles per hour is 32 dB.

2.4 THE SUCCESS OF THE FIRST-GENERATION CELLULAR SYSTEM

The cellular system development went on for almost 20 years, from 1964 to 1983. The system design procedures were taken from research to system to switching to development. The Bell system had a slogan in the 1960s saying that all of Bell

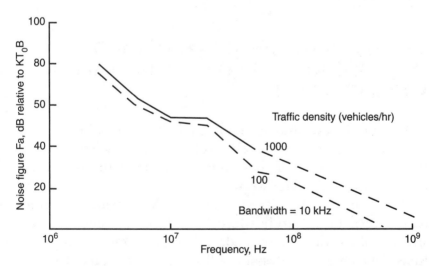

Figure 2.3. Average automotive traffic noise, F_a, as a function of frequency, bandwidth 10 kHz.

system's equipment should last for 40 years. It implied the quality. Therefore, the AMPS was designed for a high-quality product. To be sure the AMPS would be a high-quality system, a trial from 1974 to 1977 was carried out in Chicago with 16 cell sites made by AT&T Western Electric Company and 2000 mobiles made by three companies, Oki, Motorola, and E.F. Johnson. The trial was extended to 5000 mobiles. Many corrections, modifications, and improvements to the AMPS system came from the Chicago trial.[5] After the trial, a specification was written. Because the new specification was made for commercial products, the 5000 trial mobile units could not operate in the commercial system and became obsolete. The cost of this trial was very high. Today no company can afford to conduct such a large-scale trial. AMPS is a high-quality system, even based on today's standard. This experience tells us that if the selection of technology is right, the cost should not be the first concern. Quality is the key element to successfully deploying the system. As long as the customer's need for the system is great, the technology advancement and the volume of the product will drive the cost down. In 1984, the car phones were large and their prices were over $3000. Fifteen years later, the handsets were small and light, and the prices were close to $150.

2.5 SENDING THE SIGNALING AND INFORMATION DATA OVER THE ANALOG VOICE CHANNEL

AMPS has a clever way of sending a burst of data for the control signaling over the analog voice channel when needed. Because the energy of the human voice is in the bandwidth from 300 to 3000 Hz, we would like to find a waveform of data stream that has its spectral energy outside this voice bandwidth. Fortunately there is a waveform code of data stream called Manchester code, whose spectral energy of 10-Kbps data stream resides at 8000 Hz. Now we can place a filter at 8000 Hz. If the energy of a signal is detected at 8000

Hz, we know a burst of data is coming. We then can mute the voice channel for 100 ms. In this case we do not need to use a separate control channel for handoff. In AMPS, when the voice channel takes over the call from the control (setup) channel, all the control functions will be carried out through the voice channel. Manchester code is a unique kind of waveform code, not an error-correcting code. The 100-ms mute of voice is called Blank and Burst. This Blank and Burst function is later used in the digital system for separating the control signaling from the digital voice stream. However, when sending information data over the AMPS voice channel and during the data stream recovery in the handoff period, the data stream should be predistorted before sending through compression and preemphasis in the FM system. While receiving, the data stream will be going through deemphasis and expansion before decoding. Without predistortion, the data transmission is always a problem because the system should treat voice and data differently. However, AMPS was only designed for voice.

Even in a digital system, the voice data stream and the information data stream are treated differently. That is why the voice goes through the vocoder and the information data bypasses the vocoder. This bypass device is called interworking function (IWF).

2.6   REPETITION CODE IS NOT A BAD CODE FOR MOBILE RADIO[6]

The repetition code transmits a stream of information bits repeatedly to generate redundancy. One bit can be repeated an odd number of times. After receiving, each correct bit will be determined by the majority vote. We call this process a majority vote detection scheme.

The repetition code is generally not an efficient code for data transmissions. However, it is still used because any code applied to correct the errors through transmission has to understand the transmission medium. If stationary, the fading wave structure in the multipath medium (i.e., the mobile station traveling at a constant speed) is stationary, then the correct bit can be found based on a proper forward error correction (FEC) code. But the fading wave structure is not stationary. The fading wave frequency will vary as the mobile speed changes. Then the length of burst errors changes in real time to force the interleaving period of data stream and the FEC code capability to change accordingly. It is difficult to find an existing code to cope with this condition. As a result, the repetition code becomes a suitable code for cellular AMPS.

The data transmission in a wire-line medium often applies Acknowledge Return Request (ARQ) without using FEC. A data stream is divided into frames. Each frame contains many bits. If any one frame has errors after detection at the receiving end, the sender is asked to resend the entire frame. Sometimes this scheme can be called "error-free" transmission because the wire-line medium is so quiet that not many resends occur. On the contrary, the wireless medium is very noisy. Most frames do have errors after receiving. If only the ARQ scheme is used, every frame has to resend many times or endlessly. Under certain conditions in the wireless medium, the so-called error-free transmission may become "no" transmission. Therefore, in mobile medium, ARQ is hardly ever issued solely. Repetition code is used because of the unpredictable change of the medium generated by the variable speed of the vehicle.

2.7 THE FINDINGS OF HANDOFF DIFFICULTIES

In AMPS, the handoff process is a hard handoff (i.e., "break before make"). In this situation, the mobile unit has to receive the correct information to hand off to the right cell at the right frequency channel. Otherwise the call drop occurs. In the beginning of writing the AMPS specification, the handoff signaling had the same repetition format as the control channel (or setup channel). In the control channel, the information data is sent five times and each correct bit is decided by making a majority vote among five repeated bits received. Due to the multipath fading, sometimes the five repeats of a signal may not be enough, especially in a weak signal spot. Unfortunately, the handoff usually takes place in a weak signal spot. The first experimental results highlighted the need to increase the number of signaling repeats from five to seven due to the high error rate. We then had to reprogram the 8-bit microprocessor to implement the seven repeats algorithm. From the seven repeats result, we realized we had to go to nine repeats. After nine repeats, we went to eleven repeats and then thirteen repeats. Finally we concluded that eleven was the right number of repeats to obtain the specified error rates. It took roughly 6 months to merely find the proper number, eleven repeats, for handoffs. As a result of this patience, AMPS was proven to be an excellent, well-developed system before its deployment.

The hard handoffs are used in analog and TDMA systems. In 1989, CDMA was invented for cellular use. It uses soft handoffs (i.e., "make before break"). Although in the handoff regions, the soft handoff reduces the call drops best, it also reduces capacity compared with the capacity of non-handoff regions. Overall, the CDMA is a high-capacity system. It will be mentioned in Chap. 6. The previous mentioned three systems, analog, TDMA, and CDMA, are all Frequency Division Duplexing (FDD) systems. FDD uses a pair of spectrum bands, one for transmit and one for reception.

In the Time Division Duplexing (TDD) system, only one unpaired spectrum band is used. This system's handoff function is different; it is called a baton handoff. In the handoff region, the call continuously switches to the strong signal cell site at all times. The improvement of call drop rate would be engineered differently.

2.8   CELLULAR TERMINOLOGY IN NORTH AMERICA

2.8.1   Why Named Forward Link (FL), Not Downlink (DL)

The terms *uplink* and *downlink* are used for satellite communication. The link from satellite to earth is the downlink.

The link from earth to satellite is the uplink. In satellite communication, the satellite position is always higher than the ground base station. The terms uplink and downlink cause no confusion.

In the cellular system, the use of uplink and downlink may cause confusion. Uplink sometimes means from base station to mobile and sometimes means from mobile to base station, as shown in Fig. 2.4. In addition, an airplane can be a mobile, just like a satellite, in which case the uplink is the link from the base station to the airplane. Uplink and downlink become very confusing in terrestrial and air-to-ground communications. Therefore Bell Labs scientists dropped the terms *uplink* and *downlink* and use forward link and reverse link. In the North American system,

The forward link (FL) is base to mobile.

The reverse link (RL) is mobile to base.

In the GSM system,

Uplink (UL) is mobile base (equivalent to reverse link).

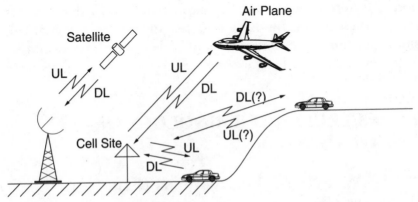

Figure 2.4. The varying uses of *uplink* (UL) and *downlink* (DL).

Downlink (DL) is base to mobile (equivalent to forward link).

In the satellite system,

Upline (UL) is earth (base) to satellite (mobile) (equivalent to forward link).

Downlink (DL) is satellite (mobile) to earth (base) (equivalent to reverse link).

As shown above, the terms *forward link* and *reverse link* have less confusion than the terms *uplink* and *downlink*.

2.8.2 WHY CELL SITE, NOT BASE STATION

The term *cell site* was coined by Frank Belcher of Bell Labs. He felt that, in 1975, the conventional base stations installed only transceivers, but the cellular base station function was more than just a base station with an added controller. The cellular base station could scan the signal strengths of the voice channels and report to the switching office, order the power controls of the mobile units, and conduct the handoff process. To distinguish the cellular base station from the existing base station at that time, the term *cell site* was used with AMPS.

2.8.3 WHY HANDOFF, NOT HANDOVER

When the feature of handoff (see Sec. 2.7) was invented in AMPS, the term *handoff* was used because it was used in football, as in to "hand off a football." *Hand over* means "to yield control of" or "to surrender control of." Therefore, *handoff* refers the call itself, whereas *handover* refers the control of the call. Either of them is okay. Because *handoff* was already used in AMPS, there is no confusion among the new engineers in the United States. In Europe, however, they did not know where the term *handoff* came from, and feeling that it was not a proper term, changed it to *handover*.

2.9 SELECTIVE FADING AND NONFADING CONDITIONS[2]

In the mobile radio environment, the multipath fading signal is created by the multireflections caused by buildings. Each frequency creates a fading signal, but different frequencies create different fading signals in the time domain. Those fading signals are called selective fading signals. The multipath fading signal only can be observed while the mobile is moving. If the mobile is at a standstill in a multipath reflected environment, the nonfading signal is observed. The signal level in this case depends on the location of the mobile. It could be a high signal or a low signal or under the fade. Different frequencies can have different received signal levels at one location. Therefore, when the mobile is at a standstill, no matter whether it is in a construction area or not, no fading signal is observed. Also, when the mobile is moving in a nonconstruction area, no fading signal is observed either.

At present, all the cellular systems use FDD, and the transmitting and the receiving channels are on different frequencies. Therefore to make a call, four frequencies are used, two for control channels and two for voice channels. When the vehicle is moving, the system operates based on four average signal strengths over their individual fading signals. These four frequencies are selective fading signals, which means that the fades do not occur at the same time (or same location) for the different frequencies as shown in Fig. 2.5. When the signal strength is averaged over a period of time, the actual fading spots are not affected. There is a constant relationship among the four average signal strength levels. However, when the vehicle is at a standstill, the four channels become four nonfading signals. The signal level of each channel should be determined by the location of the vehicle. This is called the selective fading phenomenon. There is no constant relationship among the four nonfading signal levels at any fixed location anymore. Therefore, the design of the cellular system had to consider both moving and standstill cases.

In the moving condition, the average signal strength level of any one frequency channel can represent the signal coverage of a whole area. When in a standstill condition, the signal strength level of one channel can only represent itself (not the other three channels) at the site where the vehicle is located. To make a call, all four frequency signals have to be above level S shown at D1 locations in Fig. 2.5.

2.10 THE SUCCESS OF THE ESS SYSTEM AND THE APPLICATION FOR MOBILE SWITCH

In 1950 the No. 5 crossbar switch was developed. It was a mechanical switch and performed very well. The electronic switching system (ESS) was started even before the No. 5 crossbars were fully deployed to the local switching offices.

Raymond Ketchledge at Bell Labs led a team to develop ESS. He believed that the future switch would be a software switch, not a hardware one. The added features on ESS would be very easy. Designing a special-purpose computer for the communication network would require a big drive, because the software could create a significant human-error problem. These potential errors and the integration of subrouting programs could delay the schedule and cause users to run out of patience. In this stage, engineers developing the No. 5 crossbar were ESS's rivals. However, under Ketchledge's leadership, the team never gave up. In 1967, No. 1 ESS was successfully demonstrated. He killed the No. 5 crossbar. Of course during the development stage of No. 1 ESS, Ketchledge made a lot of enemies in Bell Labs.

Later, the advantage of using ESS was proven. Although the No. 5 crossbar's quality was very good, its life was very short. In addition, the mobile switch couldn't be realized without using ESS. But with No. 1/1A ESS, the mobile switch was developed within 6 months.

Learning the software program management and control of any ESS was difficult during development. Many failures occurred. One example was ITT's System 12 switch. It was an ESS switch developed by Bell Manufacture Co. in Belgium. ITT wanted to convert its function standard protocols from European to American and sell it to large telecommunication operators in the United States. The capacity and features of System 12 as introduced were very impressive. Many telephone companies, including AT&T, had placed the orders. But the conversion from one standard to another was a mess. ITT spent 3 years and more than $2 billion on the problems in the early 1980s before finally giving up.

2.11   IN-BUILDING PROPAGATION IS THREE DIMENSIONAL[7]

For the ground vehicles, the signal propagation is two dimensional (2D) (i.e., the signal reception on the ground). When the signal is received on different floors in a building, the nature of signal reception changes. First, the signal penetrating through the walls causes a great deal of loss, roughly 15 to 28 dB, depending on the building construction, as shown in Fig. 2.6. In Chicago, no earthquakes occur and the buildings are constructed using only main frames. The signal penetration loss through buildings in Chicago is about 15 dB. In

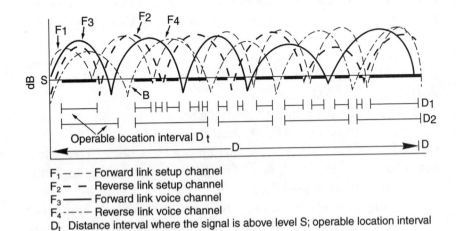

F_1 – – – Forward link setup channel
F_2 – – Reverse link setup channel
F_3 —— Forward link voice channel
F_4 –·–·– Reverse link voice channel
D_t Distance interval where the signal is above level S; operable location interval
D Total distance of Interest
D_1 Total distance that all four frequencies are above level S
D_2 Total distance that only F_4 is above level S

Figure 2.5.   An illustration of FDD with four selective fading signals.

Japan, the earthquake construction code requires structures to be built using a mesh frame. The signal penetration loss through Japanese buildings is as high as about 28 dB.

Another phenomenon is that the signal is stronger when the floor is higher above the ground. This creates a major problem. When the user is on the first floor, the signal is much weaker than for a user outside. On the sixth or ninth floor, the signal is stronger and roughly the same as the signal received on the outside ground level. And on the twentieth floor or higher the signal received is much stronger than on the outside ground. When a cellular phone is used on the twentieth or higher floor, the cell radius R increases and the rule for the cellular system required $(D/R)_s$ is violated (see Chap. 1). The mobile phones in the adjacent cochannel cell can interfere with the handset phone in the target cell. Therefore, many new techniques are used to overcome this situation. One is to dedicate 10 or more cellular spectrum channels to the in-building use. Those same 10 channels can be used on each floor because there is 20-dB signal isolation between floors, as shown in Fig. 2.6. Ten channels can generate 100 talking channels in a 10-floor building. Furthermore, the signal isolation between buildings can be 40 dB and up. The same 10 channels can be used in other buildings. The frequency reuse will be 100- or 1000-fold. It is an excellent spectrum efficiency approach. Most of all, the same handsets can be used for outdoor and in-building use.[18]

2.12 PERFORMANCE OF NTT'S VERSION OF AMPS

In 1978, NTT took AMPS specifications and modified them. The AMPS channel bandwidth of 30 kHz was changed to 25 kHz to fit into the current network at that time. The capacity did not change by reducing the bandwidth, as explained in Sec. 3.4. With the same power, the broader bandwidth (30 vs. 25 kHz) covers a greater distance. In the United States the broader bandwidth is the right choice because a vast amount of land must be covered.

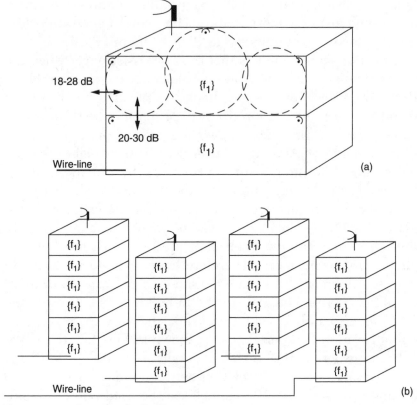

Figure 2.6. Concept of inbuilding communication.

NTT deployed the cellular system in 1979 and did not implement the diversity schemes at the base stations. The AMPS system demonstrated that diversity was a must to reduce signal fading in mobile radio signals at that time. NTT customers were not very pleased with the quality of service during this time. In 1984, U.S. cellular systems used the diversity scheme and proved its signal superiority. In 1989, NTT's new cellular high-capacity system implemented the diversity scheme both at the base station and at mobile stations (or handsets). Use of the diversity scheme became a requirement in cellular systems.

The AMPS use 10-kbps signaling speed for the control channel. NTT was using multitone frequency (MTF) at 300 bps in 1979. But the MTF at 300 bps was not suitable for mobile communications. The additional receiver noise generated while a vehicle is moving is called random FM noise. It is around 200 Hz at a high mobile speed. This low signaling rate could be interfered with by the random FM noise. That is why a low signaling rate cannot be used for call setups in a high-speed vehicle. Due to lack of many new techniques in NTT's early cellular system, the quality of the system was not satisfying to customers in 1979.

On the contrary, AMPS was deployed in 1983 with many new techniques, and the quality was very good. The customers could use it for their business, which justified the phone bills. The cellular phone markets started to take off. Although the Japanese were the first ones to deploy cellular systems, the market actually took off after systems were deployed in the United States and other parts of North America.

2.13   THE VALUE OF THE SIGNAL STRENGTH PREDICTION TOOL

The earlier signal strength prediction for the fixed-to-fixed propagation was developed by K. Bullington, R. Young, and others. After Y. Okumura published a paper in 1969, the engineers started to realize that the signal strengths received in urban and suburban areas were different due to the different effects of human-made structures. To design a cellular system for good coverage and successful handoffs, a prediction model based on the terrain configuration information was created by W. C. Y. Lee.[8] A prediction tool called Area Coverage Estimation (ACE), based on Lee's model, was developed at Bell Labs. The ACE program was later modified and renamed Advanced Mobile Design System (AMDS). This tool helped all RBOCs deploy their cellular systems between 1983 and 1985. The first cell size was about 8 miles in radius. Choosing a proper site was not easy. Without ACE, obtaining the measurement data in planning a deployed area could be a time-consuming task. Using the tool, the analysis from the predicted data could be used to determine the cell sites. *The cost and the latency of using the measurement approach were so great that the prediction tool was adopted.* The microcell prediction tool is also very important. Deploying the microcells in proper locations is critical for cost and interference issues. Many microcell prediction tools have been published.

The in-building prediction tool may not be useful because each building's structure is different. Each floor of a building has a different layout. The prediction tool has to digitize the floor space, input the structural material, and then plot out the predicted coverage of each floor. *The signal strength prediction is based on statistics, and the microprediction (in building) has only a small database to form a statistical nature. Often the predicted coverage of each floor is far off from the measured data. It is better to just make a measured run of each floor, and the coverage of the floor will be shown.* The correction or change of the antenna probes can be easily done in a building. The database can be stored for future use. Therefore there is no really commercial need for the in-building prediction tool. Instead a measured cart that can be moved around with coverage-plot capability is needed.

Commercial and academic interests are often different. The in-building prediction tool is of academic interest rather than commercial interest. However, a simple in-building prediction tool was developed for estimation purposes.[9]

2.14   COCHANNEL INTERFERENCE IS A KILLER

We know that any drug cure has its side effects. It is no different in the cellular system. We are trying to reuse the same frequency in different locations to increase the spectrum efficiency. Nevertheless, at the same time, the interference caused by many cochannel frequencies is a great problem because the terrain is not always flat and the required $(D/R)_s$ value may not be good enough. When in a large cell size, such as an 8-mile radius cell, the $(D/R)_s$ rule is much easier to apply. When the cells become small, cell site cannot be chosen properly due to the actual situation. The cochannel interference starts to rise in the system. Cell splitting is an ideal technique to increase capacity. But because of cochannel interference getting stronger, the one-to-four capacity increase by splitting a cell into four small cells cannot be observed. On the other hand, we can increase the D/R ratio to further reduce the cochannel interference. But the larger the D/R, the less the channel capacity. Therefore we try to keep the D/R ratio small by using other techniques such as sectorization, antenna down tilt, microcell frequency management, and smart antennas to reduce the cochannel interference.

The reduction of interference, cochannel and adjacent channel, is a key element in the cellular system. Using directional antennas, down-tilting the antenna, lowering the transmit power, narrowing the skirt filters, and so forth, at the transmitting end are very helpful. We may have to dream up an ideal situation for reducing interference, in which an invisible line is connected between the base station and the mobile, as shown in Fig. 2.7. We have all realized that no interference exists in the wire-line transmission between two ends. Thus, let's always work toward this ideal situation by using all new ideas and means.

2.15   COVERAGE OF 39 VERSUS 32 dBμ

The FCC has used a specified received signal strength for the coverage boundary, which was 39 dBμ (dB is mV/m) up to 1989. In Lee, 1989,[10] Sec. 7.6.6 was called "A 39-dBμ and a 31-dBμ boundary." In this section, Lee emphasized that in reality the cell boundary or the handoff is based on the voice quality, that is, C/N = 18 dB or a level above the environmental noise level of -118 dBm (i.e., -100 dBm). There is a conversion of dBm to dBμ (dB in μV/m) based on monopole antenna matching on a 50-Ω load at 850 MHz.

The level of -100 dBm is 32 dBμ. Therefore, the FCC cell boundary of 39 dBμ (or -97 dBm) is 7 dB higher

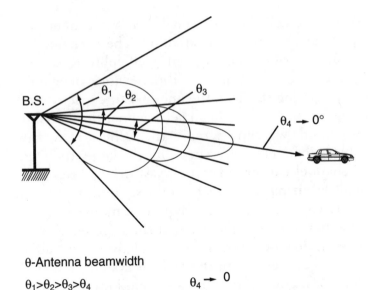

θ-Antenna beamwidth

$\theta_1 > \theta_2 > \theta_3 > \theta_4$ $\theta_4 \to 0$

(a) The best case of delivering power to the mobile unit in space by antenna

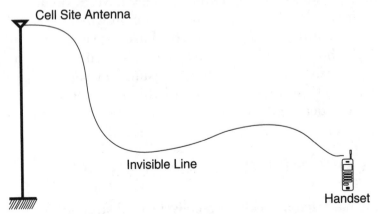

(b) The ideal situation for the wireless communication

Figure 2.7.   A thearetical ideal for reducing interference.

than the level provided by the system. Thus, a cell bound-ary of 32 dBμ (or −100 dBm) proved to be sufficient for cellular coverage. In 1992, the FCC made a rule that changed the cell boundary requirement from 39 dBμ to 32 dBμ. The two main advantages[10] of using a 32 dBμ contour

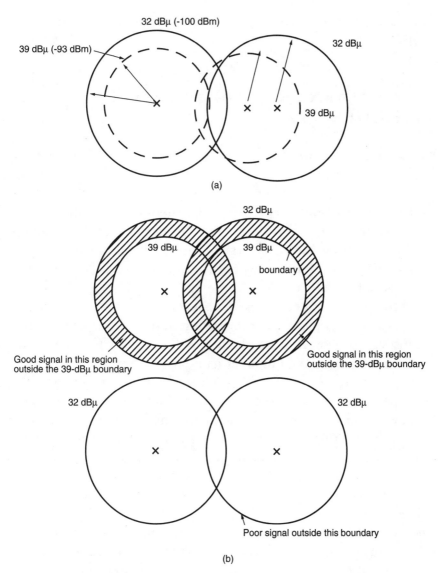

Figure 2.8.   The main advantage of a 32-dBμ contour.

(see Fig. 2.8) are (1) fewer cell sites would be needed to cover a growth area, as shown in Fig. 2.8*a*, and (2) less interference would be effected at the boundaries, as shown in Fig. 2.8*b*. A 32-dBμ boundary of a metropolitan statistical area (MSA) or a rural service area (RSA) is a proper operation, as opposed to a 39-dBμ boundary, which is an artificial value.

2.16 ADVANTAGES OF DIVERSITY SCHEMES

In a mobile environment we have found that the diversity schemes, not the directional antennas[11] (see Sec. 5.91) can reduce the multipath fading effectively. There are many diversity schemes: space diversity[12] using separated antennas, field components (or energy density) diversity using E and H field components,[13-15] polarization diversity using vertical and horizontal polarized waves,[16] path diversity,[17] and so forth. The diversity gain is not a real gain, just as the coding gain is not a real gain and the processing gain of spread spectrum is not a real gain. If the environment is free from fading and interference, all the gains stated above are unaffected by the signal. The real gains are the antenna gain, amplifier gain, and so forth. However, the diversity gain can help reduce the transmit power requirement for the same coverage, reducing the burst error rate for the signaling and data transmission and reducing the time delay spread for high-speed data. Diversity is in general used at the receiving ends, both the base station and the mobile station (or handsets). This is because the receiving diversity does not generate any more interference to the other users, but it enhances its own reception performance. Recently, transmit diversity was proposed: transmit a signal with two orthogonal components onto two transmit antennas (see Fig. 7.15). Implementing transmit diversity at the base station simplifies the handset radio. But the drawback is that the transmit diversity will split the transmit power into two for a two-antenna system. It has a 3-dB power degradation in signal transmission. Compared with the receiving diversity, there is no degradation. The transmission diversity and the polarization diversity have the same issue. The polarization diversity needs to transmit two different polarized waves from two antennas. The 3-dB transmit power degradation was the reason in 1974 to disregard the polarization diversity.

2.17 REFERENCES

1. CCIR Report 719-1 Attenuation by Atmospheric Gases, CCIR XV Plenary Assembly, Geneva 1982, Vol. V.
2. W. C. Y. Lee, *Mobile Communication Design Fundamentals,* 2d ed., New York: John Wiley, 1993. pp. 345–349
3. D. C. Cox and R. P. Leck, "Distribution of Multipath Delay Spread and Average Excess Delay for 910 MHz Urban Mobile Radio Paths," *IEEE Trans. Antenna Prop.* vol 23, March 1995, pp. 206–213.
4. A. D. Spaulding, "The Determination of Received Noise Levels from Vehicular Traffic Statistics," *IEEE Nat. Telecomm. Conf. Record.* 19D-1-7, December 1972.
5. Bell Labs, "High Capacity Mobile Telecommunications System Developmental System Reports" No. 1–No. 8 published every 3 months from March 1977 to March 1979, submitted to FCC.
6. W. C. Y. Lee, "The Advantages of Using Repetition Code in Mobile Radio Communications," *1986 IEEE Vehicular Technology Conference,* May 22, 1986, Dallas, Texas.
7. W. C. Y. Lee, *Mobile Cellular Communications, Analoged Digital System,* 2d ed., New York: McGraw-Hill, 1995, pp. 594–598.
8. C. S. Phelan, Bell Lab Patent Attorney's letter, "A New Mobile Radio Propagation Model Case 39445-7," dated March 30, 1979. (See also Exhibit 1.B)
9. W. C. Y. Lee and David J. Y. Lee, "Computer-Implemented Inbuilding Prediction Modeling for Cellular Telephone Systems," U.S. Patent Office has granted a patent.
10. W. C. Y. Lee, *Mobile Cellular Telecommunication Systems,* New York: McGraw-Hill, 1989, pp. 229–231.
11. W. C. Y. Lee, "Preliminary Investigation of Mobile Radio Signal Fading Using Directional Antennas on the Mobile Unit," *IEEE Trans. Veh. Comm.,* vol. 15, no. 2, October 1966, pp. 8–15.
12. W. C. Y. Lee, "Antenna Spacing Requirement for a Mobile Radio Base-Station Diversity," *Bell System Technical Journal,* vol. 50, July-August 1971, pp. 1859–1874.
13. E. N. Gilbert, "Energy Reception on Mobile Radio," *Bell System Technical Journal,* vol. 44, October 1965, pp. 1779–1803.
14. W. C. Y. Lee, "Statistical Analysis of the Level Crossing and Duration of Fades of the Signal from an Energy Density Mobile Radio Antenna," *Bell System Technical Journal,* vol. 46, February 1967, pp. 416–440.
15. W. C. Y. Lee, "An Energy Density Antenna Model for Independent Measurement of the Electric and Magnetic Fields," *Bell System Technical Journal,* vol. 46, September 1967, pp. 1587–1599.

16. W. C. Y. Lee and Y. S. Yeh, "Polarization Diversity System for Mobile Radio," *IEEE Trans. Comm.*, vol. 20, no. 5, October 1972, pp. 912–923.

17. A. Salmasi and K. S. Gilhousen, "On the System Design Aspects of Code Division Multiple Access Applied to Digital Cellular and Personal Communications Networks," *IEEE VTC '91 Conference Record*, St. Louis, May 19–22, 1991, pp. 57–62.

18. W.Y.C. Lee, "Inbuilding Telephone Communication System," U.S. Patent Office, No. 5,439,631, Sept. 20, 1994.

HOW TO EVALUATE A SPECTRUM—AN EFFICIENT SYSTEM

3.1 DEMAND AND CAPACITY ISSUE

In this fast-growing field, the wireless communication industry improves system performance by constantly introducing new technologies. However, end users benefit from the performance but are not concerned with technology; they only care about service features, terminal size, and cost. Voice quality and system performance must also be acceptable to them.

Voice quality and system performance are inversely proportional to service demand and system capacity, especially in wideband communication services. Higher demand in service may mean a trade-off with lower quality in the system. With the inefficient spectrum coordination by the FCC, end users will soon tire of poor quality or system performance in advanced wireless communication systems. Besides, the FCC's policy in spectrum coordination and spectrum sharing has a great impact on the demand and capacity of the markets.

FCC's system coordination charter has become less confined. The license auction winners can have a great flexibility in using their licensed spectrum. They can even sell a portion of this owned spectrum. The interference issue would affect the demand and capacity not only to its own system but to the neighboring system as well.

The spectrum-sharing policy also promotes some new entrepreners who do not have enough financial support to get the spectrum through auction, and take a low-cost approach of using the current cellular system equipment with the cellular spectrum for operating their service. The cellular spectrum is dealing constantly with demand and capacity with its own system as the number of subscribers keeps increasing. Now the cellular operators have to pay more attention and worry about the spectrum-sharing strangers.

3.2 HOW TO CALCULATE THE RADIO CAPACITY OF ANALOG CELLULAR SYSTEMS[1]

In all cellular systems, the radio capacity is measured as "the number of channels per call per MHz":

$$m = \frac{M}{K} \qquad \text{\# of channels/cell/MHz} \qquad [3.1]$$

where M is the total number of channels in 1 MHz. In FDMA or TDMA, M is a known number. In AMPS M is $395/12.5 = 31.6$ channels/MHz. The number 395 is the total voice channels of AMPS and K is the frequency reuse factor expressed in Sec. 1.3. K is related to D/R ratio where D/R in turn is related to C/I (carrier-to-interference) ratio:

$$\left(\frac{C}{I}\right) = \frac{(D/R)^4}{6} \qquad [3.2]$$

Equation [3.2] is based on six interferers, as shown in Fig. 1.1. Equation [3.2] can be used for FDMA and TDMA but not for CDMA. The required C/I will be determined by voice quality or data performance. In the analog AMPS system, $(C/I)_s$ is 63, which is 18 dB. Let the $(D/R)_s$ be based on $(C/I)_s$:

$$\left(\frac{D}{R}\right)_s = \sqrt[4]{6\left(\frac{C}{I}\right)_s} \qquad [3.3]$$

then

$$K = \frac{1}{3}\left(\frac{D}{R}\right)_s^2 = \sqrt{\frac{2}{3}\left(\frac{C}{I}\right)_s} \qquad [3.4]$$

K should be an integer[1] as 4,7,9,11, 13... and Eq. [3.1] becomes

$$m = \frac{M}{\sqrt{\frac{2}{3}(C/I)_s}} \qquad [3.5]$$

$$\text{For} \left(\frac{C}{I}\right)_s = 63$$

$$K = \sqrt{\frac{2}{3} \times 63} = 6.48 \approx 7$$

Substituting $K = 7$ into Eq. [3.1] and taking 33.3 analog channels in 1 MHz, the radio capacity m can be obtained as

$$m = \frac{33.3}{K} = \frac{33.3}{7} = 4.7 \text{ channels/cell/MHz}$$

M in Eq. [3.5] can be defined as the number of traffic channels in 1 MHz:

$$M = \frac{B_t}{B_c} = 1 \text{ MHz}/B_c \qquad [3.6]$$

Then Eq. [3.5] can also be expressed as

$$m = \frac{1 \text{ MHz}}{B_c \sqrt{\frac{2}{3}\left(\frac{C}{I}\right)_s}} \qquad [3.7]$$

where the value of $(C/I)_s$ expressed in Eq. [3.7] is in a linear value, not decibels.

3.3   WHY FM, NOT AM OR THE DIGITAL SYSTEM, WAS CHOSEN IN THE 1970S

Frequency modulation (FM) was chosen instead of analog modulation (AM) because the requirement of the modulation scheme for mobile radio signal should be the constant envelope modulation. FM is a constant envelope modulation. As we can see in Fig 3.1, a plain carrier frequency is contaminated when transmitting through the mobile radio medium. The envelope (amplitude) of the carrier is disturbed by the multipath fading.

The frequency of the carrier is also disturbed by the random FM noise introduced by the medium. The frequency jitters (0 to 200 Hz) by random FM (RFM) noise are caused by the motion of mobile, usually much less than the human voice frequency (3000 Hz) and can be neglected. Because the envelope of the carrier is always disturbed through the mobile radio

The carrier before sending

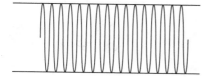

After reception

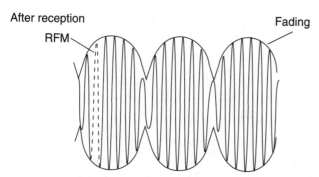

Figure 3.1. Diagram showing a plain carrier frequency (*top*) and its contamination in the mobile radio medium (*bottom*).

medium, we cannot modulate the envelope of the carrier with the information signal before sending through the medium. Otherwise, when we received the mobile radio signal at the mobile unit, we cannot easily distinguish which part of the signal is the information signal and which part is disturbed due to the multipath medium. Therefore FM is the ideal modulation because the information signal is modulated on the frequency of the carrier, not the envelope, as shown in Fig. 3.2.

The digital system wasn't chosen in the 1970s because digital technology was very expensive. In the 1970s, digital voice research was carried out at Bell Labs for the mobile radio system. The digital voice for the wireless, using modified pulse code modulation (PCM), was invented. The data rate of PCM was 64 kbps. To carry a PCM channel, the channel bandwidth has to be at least 64 kHz, which is too broad for the mobile radio channel. Therefore, differential PCM (DPCM) and delta modulation (DM) were introduced. The experimental activity for 32 and 20 kbps voice code was taking place. The voice quality was good at that time; nevertheless the digital technology was not yet mature and the cost was high. Even though

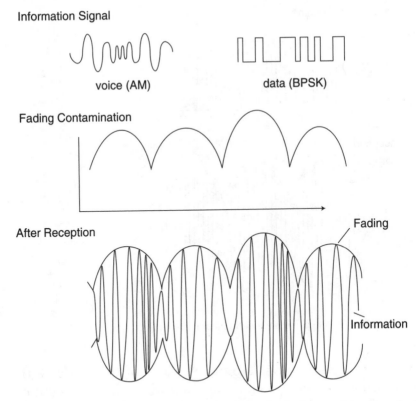

Information Signal

voice (AM)   data (BPSK)

Fading Contamination

After Reception   Fading

                                            Information

Figure 3.2.   Use of FM for the information signal, which is modulated on the frequency of the carrier, *not* the envelope.

the engineers knew the digital system should be the proper system for mobile radio, the time was not right. Then 10 years later, the European industry took the lead, developing their digital system, called GSM. It became the first second generation of the cellular system.

3.4   WHY NARROWBAND CHANNELS (SINGLE SIDEBAND) DO NOT PROVIDE HIGH CAPACITY

In 1985, the cellular industry had a group of manufacturing companies and many universities discuss why AMPS had to use a 30-kHz bandwidth for a voice channel. A 30-

kHz FM channel can be divided into six 5-kHz high side-band (SSB) channels. Then the channel capacity of the SSB system is 6 times the capacity of the FM system. In principle SSB is the right choice because it provides the most spectrum-efficient modulation. A line-of-sight communication system, such as microwave links or satellite communication, would benefit from using SSB. But in the mobile radio medium, SSB is not a suitable modulation for the following two reasons.

1. SSB is an amplitude modulation, and the ideal modulation for the mobile radio medium is the constant envelope modulation.
2. SSB is not proven to be the best spectrum-efficiency modulation in a cellular system. On August 2, 1985, Lee was invited by the FCC to give a tutorial talk on spectrum efficiency—a comparison between FM and SSB in cellular mobile systems[2] (see Exhibit 3.A).

The reason that SSB modulation is not the best spectrum-efficiency modulation is as follows: To achieve a good voice-quality channel, two related parameters are used, the channel bandwidth B_c and the required carrier to interference ratio $(C/I)_s$. When B_c is reduced, the required $(C/I)_s$ increase. Because in the analog system the signal waveform structure is unchanged, the relationship between B_c and $(C/I)_s$ is constant:

$$\left(\frac{C}{I}\right)_s \cdot B_c^2 = \text{constant} \qquad [3.8]$$

Equation [3.8] can be obtained from Eq. [3.7]. It is a theoretical formula. The voice quality for each system would be judged by human ears. For SSB signal, when its $B_c = 5$ kHz, its $(C/I)_s$ for a toll-quality voice needs to be 38 dB. A comparison of a 5-kHz SSB signal with a 30-kHz FM signal in 1-MHz spectrum band is shown in Table 3.1.

PUBLIC NOTICE

5879

FEDERAL COMMUNICATIONS COMMISSION
1919 M STREET N.W.
WASHINGTON, D.C. 20554

News media information 202/254-7674.   Recorded listing of releases and texts 202/632-0002.

July 18, 1985

Tutorial on Spectrum Efficiency A Comparison Between FM And SSB in Cellular
Mobile Systems

The Office of Science & Technology will sponsor a tutorial on August 2, at
2:00 p.m. in the Commission Meeting Room 856, 1919 M Street N.W., Washington,
DC on "Spectrum Efficiency: A Comparison Between FM and SSB in the Mobile
Cellular System." The tutorial will be given by Dr. William C.Y. Lee, Vice
President for Technology Planning and Development at the Pac Tel Mobile
Companies.

The concept of the overall spectrum efficiency includes more than just the
nominal bandwidth of the Signal. Coverage, co-channel reuse distance,
adjacent channel characteristics and perhaps other factors as well enter the
equations. Dr. Lee will explore these factors, particularly as they relate to
voice quality mobile cellular radio systems, but the concept is useful for
non-cellular systems as well.

Dr. Lee is particularly qualified to address the question of cellular
efficiency for he created the UHF propagation model used in Bell Laboratories
early development of its cellular technology. He holds a BS degree from the
Chinese Naval Academy, and MS and Ph.D degree in electrical engineering from
Ohio State. Dr. Lee is the author of Mobile Communications Engineering
published in 1982 and has a forthcoming book Mobile Communications Design
Fundamentals.

All FCC employees and the general public are invited to attend.

– FCC –

Exhibit 3.A. FCC Public Notice, "Tutorial on Spectrum Efficiency."

From Table 3.1, we have found that the radio capacity of the SSB system is less than FM because the cochannel cell separation requirement in terms of cell radius R for the FM system is only 4.6 R whereas the SSB system is 14 R. An FM system has to have a cluster of 7 cells, whereas an SSB system has to have a cluster of 64 cells. In an FM system, each cell has to have a different set of frequencies. Thus, there are $33.3/7 = 4.15$ channels per cell per MHz. In an SSB system, each cell has to have a different set of frequencies; therefore $200/64 = 3.12$ channels per cell per MHz. In this case, the capacity of SSB is less than that of FM; compare 3.12 with 4.75, shown in Table 3.1.

Table 3.1

	M (# OF TRAFFIC CHANNELS)	$(C/I)_s$ (dB)	K	$(D/R)_s$	M
FM	33.3	18	7	4.6	4.75
SSB	200	38	64	14	3.12

3.5 HOW TO CALCULATE THE RADIO CAPACITY OF A DIGITAL CELLULAR

To select a good digital system, the cellular industry needs to find a formula and use it to calculate the capacity of digital systems. On August 20, 1987, the FCC Office of Science and Technology (OST) presented a tutorial seminar on the "future of cellular radio."[3] One system operator (Pactel) and three major manufacturers (AT&T, Ericsson, and Motorola) were invited (see Exhibit 3.B). W. C. Y. Lee from Pactel indicated that you should "know the rule before you play the game." He described his newly found formula and demonstrated its use for selecting a good digital system in either multiple access, FDMA or TDMA:

1. Use of Eq. [3.1]. It was the newly found formula in 1987.

$$m = \frac{1 \text{ MHz}}{B_c \sqrt{\dfrac{2}{3}\left(\dfrac{C}{I}\right)_s}} \tag{3.9}$$

In each candidate system the channel bandwidth B_c is given. The $(C/I)_s$ can be obtained by setting the voice quality to the standard accepted level, usually at 4 at the mean opinion score (MOS) level. Now every candidate system has number of B_c and $(C/I)_s$.

2. Because every system has its own B_c, which is different from others, a normalization process is needed. The normalization

RECEIVED
AUG 26 1987
Ansd..........

FEDERAL COMMUNICATIONS COMMISSION
1919 M STREET N.W.
WASHINGTON, D.C. 20554   4521

News media information 202/632-5050.   Recorded listing of releases and texts 202/632-0002.

August 20, 1987

THE OFFICE OF ENGINEERING AND TECHNOLOGY PRESENTS A TUTORIAL ON

The Future of Cellular Radio

The Office of Engineering and Technology is pleased to present a tutorial on "The Future of Cellular Radio" on September 2, 1987, at 1:30 p.m. in the Commission Meeting Room, Room 856, 1919 M Street, N.W. The tutorial panel will consist of Dr. William C. Y. Lee of PacTel Mobile, Mr. Joran Hoff of Ericsson Radio Systems, and Mr. Jesse E. Russell of AT&T Bell Laboratories.

Dr. Lee, who delivered an earlier tutorial on "Cellular System Efficiency", is Vice-President for Technology Planning and Development at PacTel Mobile. He has a bachelor's degree from Taiwan Naval Academy, and Master's and Doctorate degrees from Ohio State University. He developed the UHF propagation model for the Advanced Mobile Phone System while at Bell Labs.

Mr Hoff is currently the Director of Cellular Systems Research and Development (Worldwide) for Ericsson Radio Systems. He has held that position since 1982. Prior to 1982, Mr. Hoff held a variety of positions with Ericsson Radio Systems primarily related to systems engineering. He has a Master's Degree in Electronics and Engineering from the Chalmers Institute of Technology in Gothenbureg, Sweden.

Mr. Russell was named Head the Cellular Software Design Department at AT&T Bell Laboratories in 1984. He has been responsible for developing computer programs that control the operations of cellular base stations. Additionally, he has responsibility for the development of new hardware and software enhancements for cellular base stations. He holds the Bachelor of Science Degree in Electrical Engineering from Tennessee State University and the Master of Science Degree in Electrical Engineering from Stanford.

The members of the panel will discuss the future of cellular radio from each of their unique viewpoints. Mr. Hoff has been directing Ericsson's efforts in a digital time division multiple access (TDMA) system similar to that chosen for the future Pan-European digital cellular system. The AT&T Bell Laboratories has recently shown a frequency division multiple access (FDMA) digital concept that would fit into the current cellular spectrum. Dr. Lee, in his tutorial cited above and several articles in the professional press, has described the benefits of wideband digital technology.

The public is invited to attend.

― FCC ―

Exhibit 3.B.   FCC Public Notice, "The Future of Cellular Radio."

formula is stated in Eq. [3.8]. $(C/I)_{SN}$ is the normalized (C/I) and B_{CN} is the normalized channel bandwidth:

$$\left(\frac{C}{I}\right)_s B_c^2 \;=\; \left(\frac{C}{I}\right)_{SN} \cdot B_{CN}^2 \qquad\qquad [3.10]$$

3. List all required new $(C/I)_{SN}$ after normalization with B_{CN}:

System A $\quad (C/I)_{SN1}$

System B $\quad (C/I)_{SN2}$

System C $\quad (C/I)_{SN3}$

4. Compare all required $(C/I)_{SN}$ after normalization. If we find that the value of $(C/I)_{SN2}$ such that

$$(C/I)_{SN2} < (C/I)_{SN1} < (C/I)_{SN3}$$

then based on Eq. [3.9], system B has the highest radio capacity and is the most spectrum-efficient system of the three.

3.6 THE REQUIREMENT OF A DIGITAL SYSTEM FROM ARTS

By 1987, cellular operators already realized that the analog AMPS could not meet the market's rapid growth. Therefore the Cellular Telecommunications Industry Association (CTIA) formed a subcommittee called Advanced Radio Technology Subcommittee (ARTS). This subcommittee had 13 members from 13 volunteer operators led by W. Lee (Pactel) and Dennis Rucker (Ameritech). The goal was to find a digital system for high capacity. The User's Preferred Requirement (UPR) was generated from ARTS:

1. The capacity needed to be 10 times AMPS
2. The suggestion by W. Lee of having dual-mode handset was introduced in the Requirement. This dual-mode handset idea appeared first in the cellular industry and was opposed by the vendors.

3. The new digital system was scheduled to be introduced to the market in 1990.

4. The digital system would first be used for voice only, not for data transmission. This came from a market survey paid for by the CTIA and carried out by Booth Alan Hamilton in 1987.

5. There would be no change for existing setup channels, using the analog setup channels to connect the call for both analog and digital voice channels. Therefore no digital setup channels were needed. It was the simplest way to speed up the developing time.

6. It would coexist with the analog spectrum. The FCC could not allocate a new frequency band for the digital cellular. Then the coexistence operation between AMPS and digital systems became a problem. In 1987, the Telecommunications Industry Association (TIA) standard body started to form the TR45.5 group to develop a cellular digital system and took ARTS requirements for its guiding principle.

At this time most industries believed that FDMA would be the winner for the following reasons:

1. There was a low risk to develop a FDMA digital system because the analog system was an FDMA system. The natural medium affecting the FDMA digital systems would have similar performance results as on the FM analog system. Many system designers have found that the FM analog system can be used for the FDMA digital system without searching for the unknown values of new parameters for a high-capacity TDMA system in the radio industry.

2. It could be commercially operated in 1990, which was merely 3 years later.

3. The time-delay spread in the mobile radio industry was relatively small for the low transmission rate in FDMA systems.

4. The developing cost was low.

5. The same value of $C/I = 18$ dB could be used for both analog and FDMA systems, so the same cell site is used for both systems. Then two systems could be cosites.

6. The capacity of FDMA could be higher than that of TDMA because:

   a. TDMA needs guard time between time slots. It is an additional overhead that reduces the capacity. FDMA does not need guard time.

b. FDMA does not need a guard band between channels because no adjacent channels would operate in the same sector or cell in the cellular system.

c. Handoffs of digital systems can be easily implemented in FDMA in the same way as in the analog system.

3.7 WHY THE TDMA SYSTEM WAS SELECTED FOR DIGITAL

In June 1987, AT&T had successfully demonstrated a FDMA system for digital in Chicago. The FDMA access scheme can be used for both analog and digital. Because AMPS is an FDMA system, many measured parameters in the field for the analog FDMA system can be used for the digital FDMA system. AT&T used a 10-kHz channel and a 8.3-kbps rate vocoder for voice. The quality was good when tested outside Chicago in a suburban area.

Later in 1987, the ARTS subcommittee of CTIA suggested, based on the AT&T demonstration, that the digital channel bandwidth or equivalent channel bandwidth should be 10 kHz and the C/I be about 18 dB for the new digital system.

Two manufacturing companies wanted to demonstrate their proposed systems. In late 1987, Motorola's FDMA system was demonstrated in Santa Ana, California, and Ericsson's TDMA system was demonstrated in Los Angeles. AT&T had had its demonstration earlier and felt it wasn't necessary to join this demonstration.

Going into the upcoming demonstration, Motorola felt that FDMA was the system that would be selected. The company wanted to impress the industry more by reducing the channel bandwidth from 10 to 7.5 kHz and by using its lab-tested vocoder of 6.2 kbps for voice. Motorola wanted to show better technologies in its demonstration than AT&T had.

Because Motorola's narrower bandwidth was only three-quarters of the original suggested bandwidth, and the vocoder was immature, the voice quality was not acceptable.

On the other side, Ericsson, a European company, wanted its TDMA system proposal to win. First, GSM, the European's future standard system, was TDMA and was in developing. Ericsson had better knowledge in developing TDMA. American and Japanese manufacturing companies had greater experience in developing FDMA and had settled on FDMA. The reason for selecting FDMA was discussed in Sec. 3.6. Therefore Ericsson knew that this demonstration was its only opportunity to become the American digital standard system. Ericsson was using a 30-kHz bandwidth channel for TDMA. Based on the requirement of having an equivalent slot channel of 10 kHz, the 30-kHz bandwidth should provide three time slots.

However, Ericsson said that implementing a three-slot TDMA channel could not be done in such a short time. Therefore it only implemented a two-slot TDMA channel within a 30-kHz bandwidth. The equivalent bandwidth then was 15 kHz, which was double the bandwidth of Motorola's 7.5 kHz. Because of an equivalent time-slot channel of 15 kHz, Ericsson could use its matured product, the GSM vocoder, which was 13 kbps. The 13-kbps GSM vocoder had enough room when sent over a 15-kHz time-slot channel but not over a 10-kHz time-slot channel. The following is a comparison of two system parameters:

DIGITAL SYSTEM	BANDWIDTH (KHZ)	VOCODER (KBPS)
Motorola FDMA	7.5	6.2 Lab tested
Ericsson TDMA	15	13 GSM

The voice quality of Ericsson's system while listening in the vehicle was far better than that of Motorola's system. Many listeners comparing the two systems only concentrated on the voice quality and said that TDMA's voice quality was much better. Not many executives understood why Ericsson's demonstration was better. No one even cared that it was not a fair comparison. Everyone at the time went for TDMA. Ericsson won the game.

3.8 EVALUATION OF A SPECTRUM-EFFICIENT SYSTEM FOR WLL[4]

The wireless local loop (WLL) system is a wireless fixed-point-to-fixed-point system. WLL is used for telephone services without wire-line connection. The WLL links are usually placed high above the ground to form line-of-sight (LOS) paths. Thus the links are in the LOS condition. In LOS condition, the radio propagation loss is the inverse of distance square (d^{-2}) instead of d^{-4} for the mobile radio loss. Therefore the propagation loss is less. The LOS condition would affect the system in three ways:

1. With a given transmit power, the signal can propagate a much further distance with the d^{-2} pathloss.
2. The interference from the adjacent cochannel cells will be higher with d^{-2} pathloss than with d^{-4} pathloss.
3. The required $(C/I)_s$ level of WLL for the nonmultipath fading condition is lower than that of cellular.

We are taking these three aspects into our calculation. To have a spectrum-efficient system for WLL, the frequency reuse scheme should apply as in the cellular system. Assume that there are six interferers, as shown in Fig. 3.3. The C/I received at a desired household is shown in Eq. [3.11] (with six interferers), which is different from Eq. [3.2] of a cellular system:

$$\frac{C}{I} = \frac{(D/R)^2}{6} \qquad [3.11]$$

Also, in the WLL system, the directional antennas can be easily used on both ends; thus Eq. [3.11] can be reduced to

$$\frac{C}{I} = (D/R)^2 \qquad \text{(with one interferer)} \qquad [3.12]$$

Then the radio capacity of WLL, M_w, is

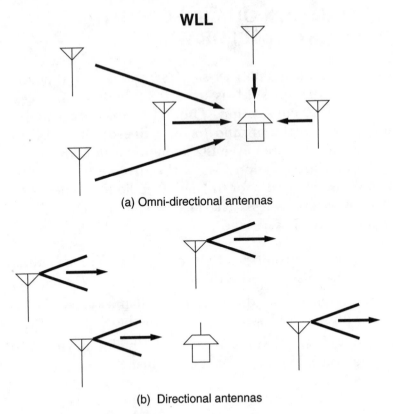

(a) Omni-directional antennas

(b) Directional antennas

Figure 3.3. WLL.

$$m_w = \frac{M}{\frac{1}{3}\left(\frac{D}{R}\right)^2} = \frac{M}{2(C/I)_s} \quad \text{(six interferers)}$$

$$= 3\frac{M}{(C/I)_s} \quad \text{(one interferer)} \quad [3.13]$$

Comparing Eq. [3.13] of WLL for one interferer with Eq. [3.5] of cellular system for six interferers, if the $(C/I)_s$ for both systems are the same,

$$\frac{m}{m_w} = \sqrt{\frac{(C/I)_s}{6}} \quad \text{(same } (C/I)_s \text{ for both systems)} \quad [3.14]$$

Because $(C/I)_s$ is always larger than 10, $m > m_w$, the radio capacity of WLL is less than that of cellular.

Fortunately the $(C/I)_s$ for both systems are different:

$(C/I)_s$ = 18 dB (=) 63 for cellular system under the multipath fading condition.

(C/I) = 10 dB (=) 10 for the WLL system under the nonmultipath fading condition.

Then, substituting these two numbers in Eqs. [3.5] and [3.13], respectively, the m/m_w ratio becomes

$$\frac{m}{m_w} = \frac{1}{3} \cdot \frac{10}{\sqrt{\frac{2}{3} \times 63}} = 0.51 \qquad [3.15]$$

The radio capacity of WLL (with one interferer) is double the radio capacity of cellular (with six interferers).

3.9 EVALUATION OF A SPECTRUM-EFFICIENT SYSTEM FOR MOBILE SATELLITE SYSTEM (MSS)[5]

The orbits used by MSS systems, GEO, MEO, and LEO, are shown in Fig. 3.4. LEO (low earth orbit) systems are starting to be deployed to the markets. In general the major specifications of using LEO system as comparing with MEO (Mid Earth Orbit) and GEO (Geosynchronous Earth Orbit) for satellite mobile communications are shown in Table 3.2. The advantages of using LEO are

1. Suitable for low power mobile handsets
2. Less latency for voice
3. Increase the radio capacity in its allocated spectrum, as shown in Fig. 3.5

TABLE 3.2 Specifications of Three Types of Satellite Communications

TYPE OF SATELLITE	ALTITUDE OF SATELLITE	AVERAGE NUMBER OF SATELLITES NEEDED FOR GLOBAL COVERAGE	BAND OF OPERATION	SUBSCRIBER UNIT POWER REQUIREMENT	DELAY	LIFE SPAN
GEO	35,000 km	3	1.6–2.5 GHz	large power	1/4 sec	15
MEO	5,000–10,000 km	10	1.6–2.5 GHz	medium power	16.5 ms–33ms	10
LEO	500–1,500 km	little = 20	little = 150 MHz	small power	2.6 ms–8 ms	5
		big = 100	big = 1.6–2.5 GHz			

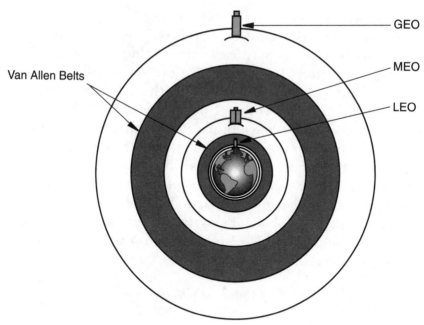

Figure 3.4. Orbits used by MSS systems.

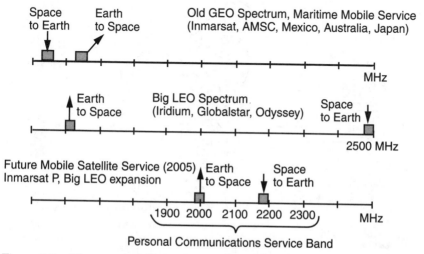

Figure 3.5. The spectrum allocated to mobile satellite systems.

The first two reasons are self-explanatory; the last needs to be addressed. The propagation pathloss is based on the free space loss because in most cases the satellite mobile link is in a LOS condition. Because there are 66 satellites in the Iridium system and 48 satellites in the Globalstar system, each satellite roughly covers 2 percent of the Earth's surface area. But due to the low altitude (900 km for Iridium system and 1400 km for Globalstar system), one revolution of each satellite around the earth takes less than 3 hours. One revolution of Globalstar satellite takes 114 minutes. Each satellite can install a multibeam antenna, and each beam can be treated as a cell. In Globalstar's system, there are 16 antenna beams, as shown in Fig. 3.6 and each beam can serve 85 channels. The radio capacity of Globalstar is

$$m = 85 \text{ channels/beam cell}$$

Because each beam may cover a 625,000-km^2 area, with a radius of 446 km, it is a huge beam cell on the earth. Then

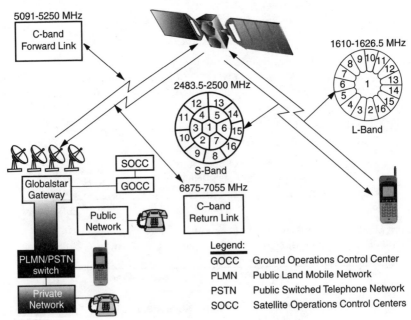

Figure 3.6. Globalstar frequency plan.

there are 0.014 channel/100 km². All the satellite mobile systems are low radio capacity systems. The LEO system has a higher radio capacity than MEO and GEO but is still not large enough to serve the urban or suburban traffic.

LEO is used to enhance cellular coverage and service in the areas where it is too expensive to provide terrestrial coverage or if some cellular customers need urgent calls and have no available channels from the cellular.

One different concept in designing the LEO system is that the cell (beam footprint) is moving very fast on the earth, as shown in Fig. 3.7. The beam cell is moving in and moving out in roughly 1 minute; a 3-minute conversation will need at least two handoffs.

3.10   INTELLIGENT MICROCELL AND ANTENNA BEAM SWITCHING CONCEPTS

As we mentioned in Sec. 2.14, the cochannel interference is a killer of system performance. A remedy is to reduce the interference by delivering the energy only to the mobile of interest in a small confined area. Two concepts were used, microcell and antenna beam switching.

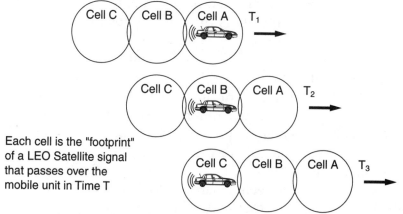

Each cell is the "footprint" of a LEO Satellite signal that passes over the mobile unit in Time T

Figure 3.7.   Cells move much faster than mobile cellular units.

3.10.1   INTELLIGENT MICROCELL CONCEPT

In 1988, Lee at Pactel patented his microcell concept, which could increase capacity 2.5 times over the AMPS without changing the AMPS's system specifications. Lee's patent was granted in 1990,[6] and an additional invention was granted in 1991.[7]

Because Pactel was a subsidiary of Pacific Telesis, which was one of seven baby Bells, Pactel had to obey the modified final judgment (MFJ) restriction issued by DOJ, discussed in Section 5.2. Following these constraints, he developed a microcell with the following advantages:

1. No change in the base station infrastructure (no need to ask vendor's help)
2. No handoffs among the microcells
3. No base station at each microcell, only a zone converter
4. Easy to deploy and take down a zone converter (same as a repeater)
5. The same voice quality as in the AMPS specification or better with a radio capacity increase of at least 2 to 2.5 times

The principle of making the intelligent microcell system work is based on the following. The frequency reuse factor K in the convention cellular system is used for both voice quality and radio capacity. In this microcell system we can divide K in two; one K for voice quality and the other K for radio capacity. The K for the voice quality is the same as for the AMPS system. However, the K for the radio capacity can be reduced from $K = 7$ to $K = 3$, which is equivalent to 2.33 times the radio capacity of AMPS.

In Fig. 3.8, we treat the microcell the zone, the zone radius R_1, and the zone separation between two active zones D_1, which obey the relation of $D_1/R_1 = 4.6$, based on $C/I = 18$ dB or $K = 7$ for voice quality. Each regular cell consists of three microcells (or zones), as shown in Fig. 3.9. The regular cell radius R and the cell separation D between two regular cells then can achieve $D/R = 3$, which yields $K = 3$. This is a 2 to 2.5 capacity increase.

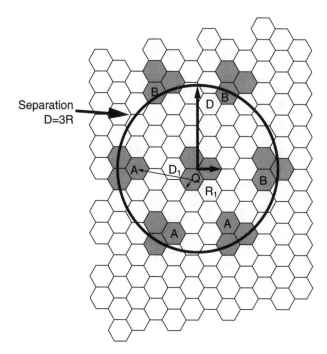

Separation
D=3R

Microcell Utilization
- For Active Zone Separation
  Zone A $D_1/R_1 = 4.6$
  Zone B $D_1/R_1 = 5.5$
- For Microcell Separation, D=3R

This Yields a K = 3

This Yields a 2 to 2.5 Capacity Increase

Figure 3.8.   Intelligent microcell capacity application.

The structure of the microcell system is also shown in Fig 3.9. The zone selector makes the zone become actively based on the measurement of the signal strength received from the mobile by the scanning receiver at the zone selector, as shown in Fig. 3.10. If there are 60 active mobile stating in three zones, the zone selector knows where those 60 active mobiles are in the three zones and sends the channel frequencies to the corresponding zones. When an active mobile moves from one zone to the adjacent zone within the regular cell, it is simply a zone switch, not a handoff. Unless the mobile is

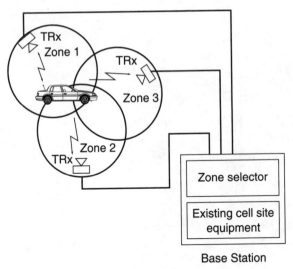

Figure 3.9. The basic microcell concept of the micro-cell system.

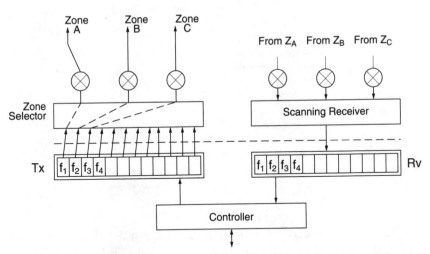

Figure 3.10. Modification of the equipment arrangement for the microcell system.

moving to another regular cell region, a handoff process is starting to take place.

The converters are in pairs. At the zone selector site there are two converters, an up converter and a down converter. The converter can either convert the frequency from 800 MHz (cellular) or 1900 MHz (PCS) to microwave (18 or 23 GHz) through wireless radio, to optical frequency through fiber, or to low frequency (around 70 MHz) through thin wire. One converter is used for transmitting and the other for receiving. Two converters, one up and one down, are also installed at the zone site. The design of the transmitting converter needs a little more attention because the signals received at the zone site from different mobiles are different due to their locations. The dynamics range of the transmitting converter at the zone site is much larger than the one at the zone selector site. The optical converter was manufactured by Allen Telecom (shown in Fig. 3.11) and the microwave converter by 3 dBm. Those converters were deployed in Los Angeles and San Diego.

Figure 3.11.   An optical converter (analog conversion), manufactured by Allen Telecomm.

3.10.2   CONCEPT OF THE SWITCHING BEAM ANTENNA APPLICATION[8]

The concept of using the switching beam antenna for radio capacity is the same as the intelligent microcell concept. The three zones becomes three antenna beams. The zone selector becomes the beam selector. Because all the antenna subbeams are collocated, there is no need to have converters. The two concepts are illustrated in Fig. 3.12. We may say the concept of using switching beam antenna is a subset of the concept of using intelligent microcell to increase radio capacity.

3.11   SEVERAL MODULATION SCHEMES FOR CAPACITY ISSUES

There are few modulation schemes for increasing capacity, which are discussed below.

3.11.1   SPACE-TIME CODING[9]

Space-time (ST) coding is a new concept. Currently the trellis code combines coding and modulation to achieve a great coding gain without sacrificing bandwidth efficiency. It uses a multilevel/phase-signaling set that signals constellations

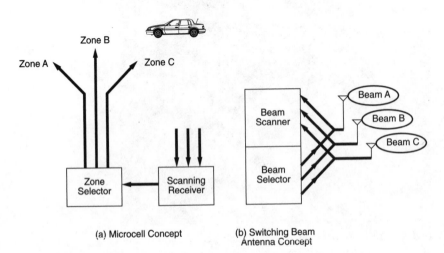

Figure 3.12.   Microcell concept vs. switching beam antenna concept.

with multiple amplitudes and multiple phases. In the ST coding for each input symbol, the space-time encoder chooses the constellation points of a modulation scheme to simultaneously transmit from each antenna so that coding and diversity gains are maximized.

Space-time coding improves the performance of cellular systems in a CDMA downlink with a limited number of antenna at the base. Each user has an individual spreading sequence (time). The unique sequence is multiplexed over a number of spatially separated antenna elements (space) to form a space-time coding. To construct a space-time code using QPSK modulation scheme, the four-state trellis code, along with the QPSK constellations S_0, S_1, S_2, S_3, defines a space-time code for a two-element transmit array, as shown in Fig. 3.13. Figure 3.14 illustrates the operation of both the code and decoder. Both the transmitter and receiver have two element antenna arrays. The fading coefficients α_{01}, α_{10}, and α_{11} are statistically independent random variables and are known a priori. Antenna T_1 will radiate the following modulation symbols $\{S_2, S_3, S_1, S_0\}$ and antenna T_2 will radiate $\{S_0, S_2, S_3, S_1\}$. The receiver antennas R_1 and R_2 will receive the fading coefficients superposed on these symbols through the medium. Then the receiver R_1 will receive $\alpha_{00} S_0 + \alpha_{10} S_1$ on antenna T_1 and $\alpha_{00} S_0 + \alpha_{11} S_1$ on antenna T_2. The received signals for each receiver antenna are depicted in Fig. 3.14. The decoder minimizes the euclidean distance between the received data and the transmitted symbols by finding the path through the trellis.[9] This code is for high-speed data transmission. This time-space code needs multiple antennas; the turbo code described in Sec. 7.12 does not. However, turbo code uses multiple processors instead of multiple antennas to decode the information quickly. Both codings promote a maximum diversity gain to the subscriber's unit. In this ST coding case, the transmitting power can be reduced because of the diversity gain providers at the mobile user. The interference level, thus, is reduced. The separation D of two cochannel cells can be smaller. The D/R can be small. The frequency reuse factor K can be smaller, and the capacity proves to be larger.

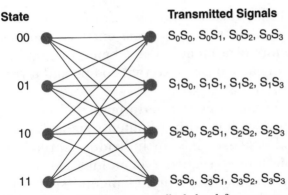

State   **Transmitted Signals**

00                    $S_0S_0, S_0S_1, S_0S_2, S_0S_3$

01                    $S_1S_0, S_1S_1, S_1S_2, S_1S_3$

10                    $S_2S_0, S_2S_1, S_2S_2, S_2S_3$

11                    $S_3S_0, S_3S_1, S_3S_2, S_3S_3$

Figure 3.13. This four-state trellis helps define a space-time code for a two-element transmit array.

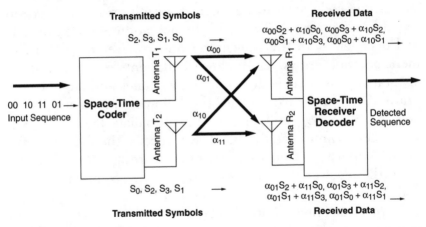

Figure 3.14. A communication system using a space-time coder.

3.11.2   APPLICATION QAM APPLICATION[10]

Quadrature amplitude modulation (QAM) can be viewed as a combination of amplitude shift keying (ASK) and phase shift keying (PSK). The ASK is not a constant envelope modulation, but PSK is. In the mobile communication system, the multipath fading contaminates the carrier envelope. Therefore the information should not be modulated on the envelope. However, QAM is a spectrum-efficient modulation. In 16 QAM, every 4 bits of information can be represented by 1 of 16 states in the transmit signal. Therefore, the required bandwidth of each channel is reduced, and the number of total serving channels increases. As a result, the radio capacity increases. Nevertheless we have to be sure that the QAM is a suitable modulation in the transmission medium of interest. The 4QAM is the same as QPSK. Figure 3.15 illustrates the 16QAM and QPSK for the noise or interference contamination. If the interference level around each state is high, the state cannot be correctively detected, and the error appears.[10] Therefore selecting 16QAM should be done very cautiously.

3.11.3   OFDM MODEM[11]

Orthogonal frequency division multiplexing (OFDM) is a modulation technique where information symbols are transmitted in parallel by applying them to a large number of orthogonal subcarriers (waveforms). The attractive advantage of using OFDM is that the modulation can be expressed in a discrete frequency domain after going through a transformation. It is a power-efficient modulation in which less spillover energy is outside the bandwidth.

The general concept is that SSB modulation is always a spectrum-efficiency modulation for voice. But it is not suitable for the multipath environment. Now the information symbols are shared by a large number of discrete frequencies. Each subfrequency carries few symbols instead of the total carrier carrying the entire symbol. For a slow rate of subfrequency carrier, the impact of symbol transmission on multiplex fading is very small. Therefore the spectrum efficiency of OFDM is high. In turn, the radio capacity increases.

3.12   VOCODERS

The vocoder is a key element in the digital system. In the mobile system, the vocoder has to be standardized. Selecting a proper vocoder depends on the type of digital system, such as FDMA, TDMA, or CDMA.

3.12.1   RPE VOCODER*

GSM has developed a regular pulse excitation (RPE) vocoder. It is a parametric-representation vocoder. A 4-kHz analog speech signal converts to a 64-kbps digital signal, then down-converts to 13 kbps through the RPE vocoder. Using a rate of 13 kbps instead of 64 kbps allows the 13-kbps data rate transmission to occur over a narrowband channel. Because the radio spectrum is a precious and limited resource, using less bandwidth per channel provides more channels within a given radio spectrum.

GSM is a TDMA system and the discontinuous transmission (DTX) mode is used. In active speech the frame is 260 bits in each 20-ms speech, and in inactive speech, the frame is 260 bits in 480 ms (24 times longer than normal mode).

3.12.2   VSELP SPEECH VOCODER†

VSELP stands for vector sum excited linear prediction. This vocoder uses a codebook to vector-quantize the excitation signal such that the computation required for the codebook search process at the receiver can be significantly reduced. It has been used for the NA-TDMA system.

3.12.3   AMR (ADAPTIVE MULTIPLE RATE)§

AMR is a new vocoder that will be used in GSM systems. AMR converts a 20-ms speech frame into an average of 80 bits. Those bits are inserted into corresponding time slots. The AMR vocoder real-time rates are multirates while sending the speech and depend on the speech profile. The average vocoder

---

*RPE-LPC Vocoder—GSM 13-kbps vocoder.

†VSELP (IS641/IS733)—8- and 13-kbps vocoder for second-generation digital systems in the United States and Japan.

§AMR (Adaptive Multiple Rate) vocoder is a new vocoder developed mainly by Ericsson and Nokia GSM and Advanced System.

rate is 4 kbps, which is half of the current 8-kbps rate. The vocoder rate drops in half and the voice channel capacity is doubled. AMR was developed for TDMA system. The DTX mode is included. It should be tested to see whether it is suitable for the CDMA system. Also AMR is not adaptive to a varied level of voice activity.

3.12.4   EVRC (ENHANCED VARIABLE RATE CODER)*

EVRC is used for cdmaOne systems because the 8-kbps VSELP voice quality is not desirable. The EVRC continuously changes its rate depending on the speech profile and the environment noise level. It is suitable for CDMA.

3.12.5   SMV (SELECTIVE MODE VOCODER)†

SMV can be used for different grades of services. It is different from the above vocoders. All regular vocoders aim at the best voice quality. This kind of vocoder aims at different grades of service specified by the customers. In this case more voice channels can be provided, and radio capacity can be further increased. There are nine candidate SMVs. The one with the best performance will be chosen and sent to the International Telecommunications Union (ITU).

3.12.6   A COMMON THIRD-GENERATION VOCODER

In developing the third-generation (3G) system, it is important to have a common 3G vocoder. Otherwise, the vocoder A in system A has to convert to vocoder B in system B. In the roaming situation vocoder B cannot work at system A even though systems A and B are both 3G systems. Therefore finding a common 3G vocoder is a must. In addition, the common audio-coder and the video-coder for 3G are needed.

---

*The EIA interium standard IS-127 for the EVRC vocoder.

†SMV (selective mode vocoder)—a kind of vocoder developed by Lucent and Conexant, etc.

3.13   HIGH DATA RATE (HDR) SYSTEM[§]

Today's cdmaOne system can be used to transmit data at a rate of 14.4 kbps. It cannot be higher because the traffic channels in one radio carrier (1.23 MHz) would transmit both voice and data. The power control is set for balancing for multiple signal reception at the base station. Therefore no one traffic channel can send more information than the others. For sending high-speed data, a system called HDR has been developed by Qualcomm and was demonstrated in November 1999 at San Diego. The system meets the increasing demand for the forseeing wireless Internet Protocol (IP) connectivity with high spectral efficiency. It allocates one or more dedicated 1.25-MHz carriers to carry IP packet data. The dedicated channel is used to send the high-rate data to different users in different time frames and with different data rates. The different data rates are based on the channel measurement of C/I. The scheduling of packet transmission takes advantage of channel diversity to multiple users. High speed is achieved using the following techniques: adaptive channel state prediction for scheduling; adaptive rate, coding, and modulation; turbo coding and decoding which transmission rate can reach the Shannon capacity; fast adaptive channel rate prediction; multiuser antenna diversity; diversity receivers, which use the optimum combination and interference cancellation; constant transmit power and variable rate; no forward link soft-handoff overhead; no fixed allocation of time slots for transmission; and enhancing medium access control. With these techniques implemented in the HDR system, the highest rate can be 2.4 Mbps. However, the high-speed rate is mostly used when the mobile is close to the base station. As the mobile moves away from the base station, the data rate is lowered. One 1.25-MHz radio carrier can send a rate of 2.4 Mbps, which is a big achievement. It is the shining light for the application of Internet using the wireless IP network.

[§]Qualcomm's HDR demonstration presentation, November 4, 1999, San Diego.

3.14   CLARITY, COVERAGE, CAPACITY, AND C/I_s [12]

Clarity, coverage, capacity, and $(C/I)_s$, the four Cs, are the key requirements for wireless communications. Also, these four Cs are mutual interactive for sending a given transmit power. The interactions are

Clarity ↑	Coverage ↓	Capacity ↓	$(C/I)_s$ ↑
Coverage ↑	Clarity ↓	Capacity ↓	$(C/I)_s$ ↓
Capacity ↑	Clarity ↓	Coverage ↓	$(C/I)_s$ ↓
$(C/I)_s$ ↑	Clarity ↑	Coverage ↓	Capacity ↓

Therefore, to keep capacity high, and also maintain the accepted quality (clarity), the cellular system deployment guidelines[12] are carried as follows. In general we convert the other three Cs to $(C/I)_s$ and find that the $(C/I)_s$ needs to be kept 18 dB for a good analog voice quality. This $(C/I)_s$ is defined as the required C/I in the system, not the measured C/I. Therefore we used the ( )$_s$ to differential the measured C/I. With $(C/I)_s = 18$ dB, we can find that the cell separation D to the radius of cell R ratio becomes 4.6. From $D/R = 4.6$, we can figure out that the analog cellular is a seven-cell cluster, or we say $K = 7$. K stands for the frequency reuse factor.

In the data transmission, the $(C/I)_s$ will be different from different digital systems. The clarity can be used for both voice (quality) and data (throughput). The interactions among four Cs still hold.

3.15   REFERENCES

1. W. C. Y. Lee, "Spectrum Efficiency Digital Cellular," *38th IEEE Vehicular Technology Conference Record*, Philadelphia, Pa., January 15–17, 1988, pp. 643–646.
2. FCC Public Notice, "Tutorial on Spectrum Efficiency—A Comparison between FM and SSB in Cellular Mobile Systems" by W. C. Y. Lee. August 2, 1985.

3. FCC Public Notice, "A Tutorial on the Future of Cellular Radio." W. C. Y. Lee delivered an earlier tutorial on "Cellular System Efficiency" followed by three presenters from three companies, Ericsson, AT&T Bell Labs, and Motorola (jointed at the time).

4. W. C. Y. Lee, "Spectrum and Theory of Wireless Local Loop Systems," *IEEE Personal Comm.*, vol. 5, February 1998, pp. 49–54.

5. W. C. Y. Lee, *Mobile Communicating Engineering, Theory and Application*, 2d ed., New York: McGraw-Hill, 1998, pp. 540–547.

6. W. C. Y. Lee, "Cellular Telephone System." U.S. patent 4,932,049, June 5, 1990.

7. W. C. Y. Lee, "Microcell System for Cellular Telephone Systems." U.S. patent 5,067,147, November 19, 1991.

8. W. C. Y. Lee, "An Optimum Solution for the Switching-Beam Antenna System," *Third Workshop on Smart Antenna in Wireless Mobile Communications, Conference Record*, Stanford University, July 25–26, 1996.

9. V. Tarokh, N. Seshadri, and A. R. Calderbank, "Space-Time Codes for High Data Rate Wireless Communication: Performance Criterion and Code Construction," *IEEE Trans. on Information Theory*, vol. 44, no. 2, March 1998, pp. 744–764.

10. W. C. Y. Lee, *Mobile Communication Engineering, Theory and Applications*, 2d ed., New York: McGraw-Hill, 1998, p. 305.

11. W. C. Y. Lee, *Mobile Communication Engineering, Theory and Applications*, 2d ed., New York: McGraw-Hill, 1998, pp. 306–309.

12. W. C. Y. Lee, "Future Vision for Wireless Communication," *Seventh IEEE International Symposium on Personal. Inbuilding, and Mobile Radio Communications (PIMRC '96)*, Taipei, Taiwan. ROC, October 15–18, 1996.

IMPORTANT FACTORS OF CHOOSING A NEW DIGITAL SYSTEM

4.1 DRIVING MARKETS

Communication systems need to find their market value prior to the development stage. Market surveys are important because system operators need to find out what customers need. For instance, in 1987, the CTIA subcommittee of Advanced Radio Technology (ART) was formed to select the second-generation digital system based on capacity.

Customer requests in 1987, listed by order of importance, were:

1. Good coverage
2. Good voice quality
3. No-drop calls
4. Low cost
5. Small handsets

ARTS asked a consulting firm to find out whether data service would have a market or not. ARTS selected one, Booth-Alan Hamilton (BAH), of three marketing consulting firms based on costs and deliverable time. The samples of customers chosen for the survey were very critical to the outcome of the survey. BAH's conclusion was that most customers were still only interested in voice, not data. Therefore, data service for the second generation was not a requirement. Based on this result, ARTS made the following decisions:

1. The high-capacity digital system should be deployed in the market by 1990.
2. The capacity of the digital system should be 10 times that of the present AMPS.
3. No data service requirement was needed.

Then the operators needed only to concentrate on the voice services. Later the short message services (SMS) became hot and could be handled by the circuit switches. Data service had not drawn attention until the 1990s when the Internet started to take off. In 1999, the Internet had 200 million users and the cellular/PCS had 500 million users around the world. By 2003, the cellular/PCS will exceed 1 billion users. Wireless data services will become an urgent need in the upcoming markets. The future markets will be in the wireless Internet, and the future network will be a merger of wireless and IP core networks.

4.2 HOW TO SPEED UP THE DEVELOPMENT OF A NEW DIGITAL SYSTEM

We had to know the development limitation before we could speed up the development schedule. They were as follows:

1. In 1987, the FCC did not allocate a new spectrum for the cellular digital system. However in Europe, GSM was allocated a new band (935–960 MHz downlink and 890–915 MHz uplink). To design a system in a new virgin band is relatively easier than in a shared band (spectrum).
2. The system was to be deployed in the market in 1990 (time).
3. Its capacity would be 10 times that of the current AMPS for the digital voice channels.

The following strategies sped up the development schedule:

1. Select a proper digital system
   a. The advantages of selecting FDMA were
      (1) *Low risk.* Most U.S. and Japanese companies knew how to manufacture FDMA equipment.
      (2) *Shorten the developing time.* The parameters for designing a FDMA system can be obtained from AMPS, which is also a FDMA system.
      (3) *Cosite.* If the required $(C/I)_s$ of the new system was the same as the AMPS, the cell sizes of two systems were the same. Two systems could share their cell sites.
   b. The concerns of selecting TDMA were
      (1) TDMA cannot be an ideal system for achieving high capacity. The guard time between the time slots are an additional overhead; also the high transmission rate resulting from TDMA forces the system to have an equalizer that is not a reliable device to deal with the time delay spread imposed on the fast transmission rate.

(2) The AMPS system is a FDMA system, which cannot coexist with a TDMA system easily. Spectrum sharing is a unique situation in the United States.

2. The following sped up the development schedule:

   a. *Use a two-mode mobile unit.*   The mobile unit has both AMPS and digital modes. The digital systems then can be deployed in spot areas and the AMPS mode can cover the rest. This arrangement can ease the whole cellular market area before it starts to operate.
   b. *Shared AMPS control channels for the digital system.* Because the two-mode mobile phone was used, the control channels could be shared. The digital system did not need to develop new control channels for itself. This was a great time saver.
   c. No need to develop the data (transmission) channels based on the marketing survey.

The recommendations to the standards body of TIA were

1. *Phase I Development.*   Finishing date was 1990.

   a. Share the AMPS control channels.
   b. For voice only.
2. *Phase II Development.*   Finishing date to be determined.

   a. Develop digital control channels.
   b. For voice and data.

4.3   DUAL-MODE ARGUMENT[1]

A dual-mode mobile unit was suggested for the first time by W. Lee in ARTS. A mobile unit has both AMPS and digital system modes in Development Phase I. The advantage of having a dual-mode phone was that it gave service providers the ability to deploy the digital system in the market much sooner, rather than waiting until an entire market had deployed the digital system. Also, with a dual-mode phone, the AMPS control channels could be shared with the digital system. This could potentially save a great deal of time designing a digital control channel, especially if no dual-mode phone was suggested.

But there were two parties against this suggestion from the beginning:

1. *The RSA service providers.*   Their argument was that they didn't need a digital system for their market because they had plenty of capacity from the analog system. ARTS used the following arguments in favor of the dual mode:

   a. The digital system was needed to handle the rapid growth in the MSA and the value based on the population (POP) was steadily increasing from $15/POP in 1985 to $32/POP in 1986 to $70/POP in 1987. In fact, at that time, a RSA could not make money by running the system. ARTS knew that the fast growth of cellular markets would increase all the cellular market's value, including RSA, once a new digital system was introduced to promote the cellular market. Without a new digital system, the cellular system would have no room to handle the growth capacity and the value of cellular markets would suffer.

   b. With a dual-mode phone, the customers using the digital phones in a MSA could still roam to a RSA with the analog phones. This feature could skyrocket the roaming revenue.

2. *The vendors.*   During the 1980s, ARTS had a hard time convincing the vendors to have a dual-mode phone. The vendor's argument was that

   a. Single-mode digital phones would be simpler to make.
   b. The cost of a dual-mode phone would be high, and the customers could not afford to buy them.
   c. If the single AMPS phone was manufactured continuously, who would buy the expensive dual-mode phone?

ARTS' recommendations were

1. Service providers initially need dual-mode phones only for congested calling areas so that the digital system could be deployed sooner.

2. The vendors would be mandated to manufacture only dual-mode phones until sometime in the future.

3. Because the cost of dual-mode phones could be high, the service providers would be willing to subsidize the cost of the new phones. The service providers could reduce the commission they paid to phone dealers. In 1986, Bell South had paid $800 for each customer a dealer could bring in. The savings produced by reducing the commission could subsidize the cost of new phones.

The technology has been advancing rapidly. The vendors resisted having dual-mode phones, but the preference for single-band handsets and mobile units was based on their fear of technology and cost in 1987. Who in 1987 would have believed that a dual-mode, triple-band handset would appear on the market in 1999?

4.4   CONFLICT OF INTEREST BETWEEN SERVICE PROVIDERS AND VENDORS

There are always conflicts of interest between service providers and vendors[2] when developing a new system. Service providers want to have a digital system to cover a large area with few base stations because each base station can provide a large number of channels. The network can also be so intelligent that many new features can be implemented rapidly without adding to the cost. Vendors, on the other hand, want to sell more base stations to service providers. Therefore, if the voice quality in the system is not good, rather than improving the software part, which does not benefit them much, the vendors try to persuade service providers to place all base stations closer together. In this way, the mobile signal could be stronger and the voice quality improved.

System providers would rather not listen to the vendor's request for more base stations, but they have no choice because they are obligated to satisfy the subscribers' needs. This is why under any new system development state, the service providers should be in the driver's seat. They are the ones who are taking the risk to serve the customer's needs. On the other hand, they have to know what price the customers are willing to pay to get what they want. Based on the service charge customers will pay, service providers determine whether they can ask the vendors to deliver the equipment with the required features and quality. Without the service provider's leadership, the vendors would select technology based on their intellectual property right (IPR) technology or try to avoid any other good technologies because of another party's IPR technology. Sometimes, this involves more risk, more time in development, or more cost for the product. In general, the vendors may not care about the cost of manufacturing the equipment if it is chosen to become standard. The service provider will buy the standard equipment anyway. Also, the performance of new wireless communication equipment may not be discovered until the call traffic of the operational market reached an unacceptable level in a year or so. It will be too late to change; therefore the vendors can sell a newly developed system to a service provider with a lot of promise before deploying the system. When a provider realizes the equipment's performance is undesirable, it may have already stocked up on the equipment. It will be too late to change from one vendor's equipment to another. It is very important to select a newly developed system from an honest and quality-oriented vendor. The harm of buying a low-cost, low-quality system would be felt for a long, long time. Some vendors are willing to listen to the operator's suggestions and requests after equipment is installed and act accordingly very quickly. These vendors are preferable for operators.

4.5 OPEN SYSTEM INTERFACES

In GSM system specifications, there are common air interfaces [between mobile and base transmission station (BTS)] called the Um interface, A interface [between BTS and base station controller (BSC)], and A^+ interface (between BSC and MSC), as shown in Fig. 4.1. Because of these open system interfaces, the service providers can buy manufacturer A's

BTS, connect to manufacturer B's BSC, and connect to manufacturer C's MSC. These open system interfaces open up a fair competition among the vendors and give the service providers a choice to change from one vendor's equipment to another's based on its cost and the quality. As a result, the end customers will benefit from the low-cost service from the open system interfaces, and quality cellular markets can be grown very fast. It may be the reason why GSM's markets grew so rapidly. In the AMPS, the common air interface (CAI) is the only standard interface. There is no standard A interface or A^+ interface. Because AMPS was invented by AT&T solely in the 1970s, AT&T did not think there was need to have an A or A^+ interface. Later, Judge Green, especially appointed by DOJ to divest AT&T, banned AT&T from operating a cellular market. The seven new RBOCs started to run the cellular markets regionally and used AT&T's equipment. AT&T had additional reasons not to standardize the A or A^+ interface. In the 1980s, cellular vendors other than AT&T, such as Motorola, Ericsson, NEC, and Northern Telecom, competed very hard to get into start-up markets. The strategy for them was to win the start-up markets with a low bid in price. Once vendor A's equipment was deployed in the market, the market would be vendor A's equipment market for the future.

When the new digital system was introduced, all major vendors felt the open system interfaces would not benefit from their business; rather they feared that many new small vendors could be encouraged to compete with them and replace their equipment components. Therefore, they resisted this open system interface policy.

4.6   HOW TO DEVELOP A GOOD STANDARD SPECIFICATION

Before the 1980s, the standard specifications of all the newly developed systems were always carefully written to be sure that the system equipment manufactured based on the standard specification would be workable in the field without any

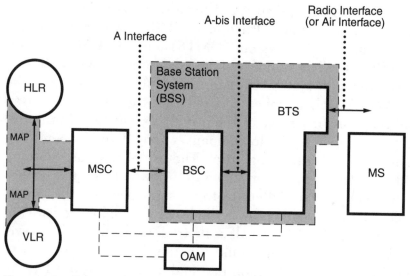

Figure 4.1.   Functional architecture and principal interfaces.

major revision shortly after deployment. Therefore, vendors prefer to manufacture new system equipment as quickly as possible to take over the market share without worrying that revisions of the standard might be issued after their products were ready to sell.

For instance, the AMP's specification was drawn up in 1979 after 15 years in the development stage, going from research to system design to switch design to commercial development. Every key parameter needed to be researched for the best of the system operation. Determining the number of repeats for the handoff on the forward link took 6 months (see Sec. 2.7). An extensive trial took place in Chicago[3] with 16 base stations and 2000 mobile units (later increased to 5000 mobile units). The preliminary specification was changed based on many findings during the trial. Once the final specification was made, it became the standard. All the 5000 trial mobile units that had been operable in the trial system were obsolete. In the 1970s, AT&T was willing to spend a huge amount of money to provide an outstanding system. No company is likely to do the same in the future.

In the 1980s, the standard for a new system was carried by a standards body. Europe used the universal mobile telecommunication system (UMTS) and the United States used the TIA. The standards body was formed from many different vendors, system operators, and universities. The self-interest of each member ensured that the new system was one that satisfied most of them but was not necessarily the best one. No one single company was willing to spend the money to carry out a huge trial. The use of few base stations and few mobile units for a trial of a newly developed system could not find the killers to the system because the small-scale trial could not provide high-capacity conditions. This was why GSM, with a small-scale trial, went through 35 revisions before the system could function in a commercial system. Writing a specification first and making revisions later is not the best practice. But because there are no strong leaders (like AT&T in the past) willing to spend the money and be patient enough to develop a well-designed system that doesn't need any critical revisions, we have to adapt the GSM development model and go through many revisions before developing a standard specification.

The service providers should sit in the drivers' seat, regardless of the development model, and set up the requirements for the new systems so the vendors can design and implement equipment accordingly.

4.7   THE FAILURE OF IS-54

In 1989, a group of company engineers working under John Stupka, the CTIA technical committee chairman, developed the North American TDMA specification. Before writing the TDMA specification, experimental information needed to be gathered for designing a high-capacity TDMA system. Unfortunately, when the team set up a date on which to finish the specification, they realized there wasn't enough time to gather the experimental data. Writing a quality system specification requires the efforts of a mixed team of experienced veterans who were involved in writing the earlier specification and new engineers who have knowledge of new system technologies. The values of each parameter in the specification needs to be measured, simulated, determined by trials, or validated. We can speed the process along, but we cannot rush the development process. Most analog and FDMA system design data are not applicable to the TDMA system. While the engineers were gathered together in a hotel to write the TDMA specification, they could only estimate most parameter values based on the GSM system, because GSM was a TDMA system. However, GSM was not the high-capacity system the group wanted to design. The specification for the North American digital system was completed in 4 months—the shortest time in history. The specification went to TIA's standards body and was assigned a number, IS-54.[4] Nevertheless, without the real measured data, the IS-54 was not a workable specification. This resulted in the failure of IS-54. Later the IS-54 specification was modified because major changes were needed. The IS-54 was changed to IS-136.[5] At least IS-136 is a workable system specification.

4.8   THE GOVERNMENT'S ROLE

With rapid communication developments in the cellular industry, the government should play a key role in guiding the cellular industry in the right direction.

4.8.1   AUCTION POLICY

The government claimed that by adopting the auction policy in 1996, the previous awkward situations that occurred during the lottery policy period between 1983–1989 were corrected. As an example, housewives and medical doctors won the lotteries and then resold the licenses they won at a substantial gain. Because the lottery policy had a disadvantage, and no other countries followed the U.S. lottery policy, why not return to the earlier selection practice where licenses were selected based on three requirements: technology competence, financial strength, and good service for public interest? All other advanced countries were practicing this method in 2G systems. The auction policy may also have created some controversial issues.

The spectrum is like air—a universal commodity. If we pollute the air, people will die. If we pollute the spectrum, wireless communications will die. The government should have regulations to protect "clean air" and "spectrum noninterfering," but the government does not own the air or the spectrum. In 1996, the government sold unowned spectrum and generated income through the auction process, which meant passing ownership of unowned spectrum to the private sector. Because the government became a profit center, it then had no power to discipline its customers, who are the auction winners. The winners could again sell a portion of their spectrum to others. The FCC duty of being a spectrum coordinator among the system operators has been lost.

In the future, the auction winners can either pass the cost of the auction to end users or deliver poor service. The auction winners can always blame the government for putting such a big burden (the auction cost) on them before they invested capital in the licensed business. This is why, in the past, advanced countries preferred to tax the service providers afterward based on their earning profits. The auction may have placed the licensees under greater financial pressure or pressed them toward bankruptcy, as happened in the PCS/C band auction. In the PCS A and B band auctions, the government received about $7 billion from the total spectrum of 60 MHz. These auctions were for a one-time payment. In the C band auctions, the payment policy changed. The winners could pay their auction fees with a downpayment, spreading the rest over 3 to 5 years. The bidding prices were very high because of the advantage of delayed payment. The government received about $10 billion from a spectrum of merely 30 MHz. However, not too long after the auction, Gateway filed bankruptcy. Then the Nextwave followed. This shows that many businesses gambled on the future. As a result, many investors, big and small, lost money, not from running the business but from the clever way that the government played and took them.

On the other hand, the auction policy stimulates the economy and creates more jobs and more high-technology opportunities. Therefore, in general, the auction policy may not be bad for the developed countries. In the developing countries, the government receives the auction money, a big portion coming from foreign investors. But the infrastructure equipment has to be purchased from the foreign countries. The cost of services will be high. Unless the average household income is high, the government may need to find a way to wisely use the auction money to stimulate the economy.

If the government thinks that the auction money comes at no cost, it is wrong. The government should be cautioned not to use auction money before the licensees' systems are built. Businesspeople may someday ask the government to return the auction money if a business is not going well. Remember, the government can never be a profit center. Nevertheless, the European countries followed suit this time for the 3G licenses. The UK government received US$ 35 billion from the auction of 120 MHz spectrum.

4.8.2   STANDARD SETTING AND SPECTRUM COORDINATION

The FCC did not want to be involved in standards setting in the past years. In broadband PCS, there were four systems and six licensed bands on the markets. This meant that any one of four systems could be operated in one of six bands in one service area that is in the same geographical area. The consequences are as follows:

1. The service provider would worry about interference from other systems in one area. In the past, one system was always standardized to one service. For instance, AMPS was used for U.S. cellular phone service. In 1992, for the first time multiple systems could be operated in the same service. Thus, the old rules of spectrum coordination could not be followed or enforced.

2. End users will pay a high price if terminal manufacturing produces a low volume for use in only one of four systems. Of course, no roaming can be done among the four systems.

3. The FCC may be unable to solve the disputes regarding spectrum interference among the different service providers. The auction process already disturbs the FCC's duty of being spectrum coordinator.

Without the FCC as an effective entity in disciplining spectrum usage, the future of wireless communication could be in a situation just like vehicles jamming an intersection in gridlock.

4.8.3   THE RISK OF A NEWLY DEVELOPED SYSTEM

A large risk is taken in the development of a new system. Both vendors and service providers will go through a learning curve. Therefore, if one system proves to be workable and meets the need, this system should be adapted to many services, such as cellular, PCS, mobile satellite, and WLL. Then end users can use one terminal to operate in different services.

4.9   DEBATE AT THE DENVER CONFERENCE

In August 1987, right after the demonstration of two systems (see Sec. 3.7) FDMA and TDMA was completed, a debating conference was held in Denver, Colorado. There were two debating teams. One team consisted of AT&T, Motorola, and NEC defending FDMA, and another team consisted of Ericsson and Northern Telecom defending TDMA. No operators were allowed in the debating teams but could sit quietly in the audience. No floor questions were allowed either. The debating conference was organized by John Stupka. TDMA was favored based on the earlier demonstration of two systems, FDMA and TDMA, in Santa Ana and Los Angeles, respectively. Although it was an unfair comparison, the debating could not be any different. After the debate, a survey was conducted. Every company had one vote, even a one-person company. As long as a company paid the $800 membership fee, it could be a member of CTIA and had one vote. Some company representatives did not know the technology and asked their alliance company whether to vote for TDMA or FDMA. The answer was TDMA. The result was that 16 voted for FDMA and 37 voted for TDMA. Thus, the TDMA became the second-generation cellular system.

It was very improper to vote on technology, because technology can be proven by theory and performance. It was later proven that the voting to select the technology was improper. As we have recognized, only those political issues that are not distinguished as exactly right or exactly wrong need the vote process. It is especially wrong when the voting process is based on one vote for one company whether it had 30,000, 300, or 3 employees. This process could be very easily mismanaged by the interested companies. Some companies that joined at the last minute did not know what FDMA and TDMA really meant but voted on them anyway. The companies that knew the technologies at that time were Ericsson, Nortel, AT&T, Motorola, and NTT. Among the five companies, Ericsson and Nortel promoted TDMA and the rest promoted FDMA. Therefore the voting result was a bit of a surprise.

4.10   DEBATE ABOUT SELECTING VOCODERS

There were 10 candidates for the North American digital system (2G) vocoder submitted to the standards body. Among the 10, AT&T's and Motorola's vocoders were in the code excited linear prediction (CELP) family and Ericsson's vocoder used regular pulse excited linear prediction code (RPE-LPC), which was similar to the one used in the GSM system. Because the equivalent bandwidth of North America's TDMA was 10 kHz, the vocoder rate had to be around 8 kbps to fit in the 10-kHz bandwidth. The CELP was an advanced vocoder at that time. The question of who was qualified to evaluate those vocoders and choose the winner was raised. No one wanted to use a U.S. evaluation company. The 10 candidates were afraid of the different interests and bias from the U.S. evaluation companies. They went to Canadian evaluating companies, Bell Northern Research (BNR) and MPT. There were many criteria for the listeners' use to score the vocoder candidates under different mobile speed from 0 to 60 mph and under different multipath conditions. Each listener generated a score sheet using the circuit-merit (CM) number on each condition shown in the following table:

CM	SCORE	QUALITY SCALE
5	5	Excellent (speech perfectly understandable)
4	4	Good (speech easily understandable, some noise)
3	3	Fair (speech understandable with a slight effort, occasional repetitions needed)
2	2	Poor (speech understandable only with considerable effort, frequent repetitions needed)
1	1	Unsatisfactory (speech not understandable)

Then the average CM scores were obtained from all the listeners. This averaged score was called the mean opinion score (MOS). Usually the toll-quality voice is around MOS $\geq$ 4. Unfortunately, the MOS of both AT&T's and Motorola's vocoder was below 4, around 3.3. Motorola's score was a little bit higher than AT&T's based on this subjective evaluation, although the CELP was first developed by AT&T. Thus Motorola's vector sum excited linear prediction (VSELP) model was chosen. Later AT&T said that it sent a wrong version to the evaluating companies by mistake. But the decision could not be reversed.

4.11   GLOBAL HARMONIZATION EFFORT[6]

The global third-generation (3G) team was set up by the ITU in June 1997. At this time the U.S. companies were developing a 3G system called cdma2000 and submitted it to the ITU. The European Telecommunication Standard Institute (ETSI) studied five different systems: TDMA/sync., TDMA/async., CDMA, Orthogonal Frequency Division Multiplexing (OFDM), and Opportunity Driven Multiple Access (ODMA). In January 1988, ETSI selected CDMA as the 3G system of choice.

Ericcson and NTT submitted their versions of 3G systems to the ITU. These proposals represented the contribution from their corresponding members, Europe and Japan, respectively. Other ITU members also submitted proposals to ITU for review. In all, 13 proposals were submitted, as shown in Fig. 4.2. Among them, eight were wideband CDMA-based proposals. Those eight CDMA system proposals were competing for the 3G standard. Among the eight CDMA proposals, some were TDD mode, some used direct spread (DS), and some were multicarrier. The 13 modes are shown in Fig. 4.3.

The convergence approach to converge 13 modes into 1 was impossible at that time. No one wanted to make concessions from their original proposal. Then another approach to

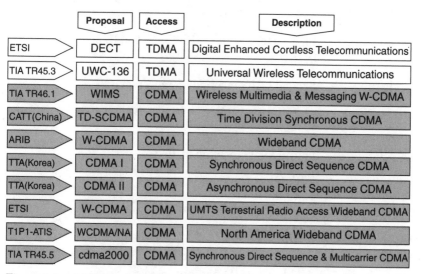

	Proposal	Access	Description
ETSI	DECT	TDMA	Digital Enhanced Cordless Telecommunications
TIA TR45.3	UWC-136	TDMA	Universal Wireless Telecommunications
TIA TR46.1	WIMS	CDMA	Wireless Multimedia & Messaging W-CDMA
CATT(China)	TD-SCDMA	CDMA	Time Division Synchronous CDMA
ARIB	W-CDMA	CDMA	Wideband CDMA
TTA(Korea)	CDMA I	CDMA	Synchronous Direct Sequence CDMA
TTA(Korea)	CDMA II	CDMA	Asynchronous Direct Sequence CDMA
ETSI	W-CDMA	CDMA	UMTS Terrestrial Radio Access Wideband CDMA
T1P1-ATIS	WCDMA/NA	CDMA	North America Wideband CDMA
TIA TR45.5	cdma2000	CDMA	Synchronous Direct Sequence & Multicarrier CDMA

Figure 4.2. 10 ITU submissions, eight are wideband CDMA-based.

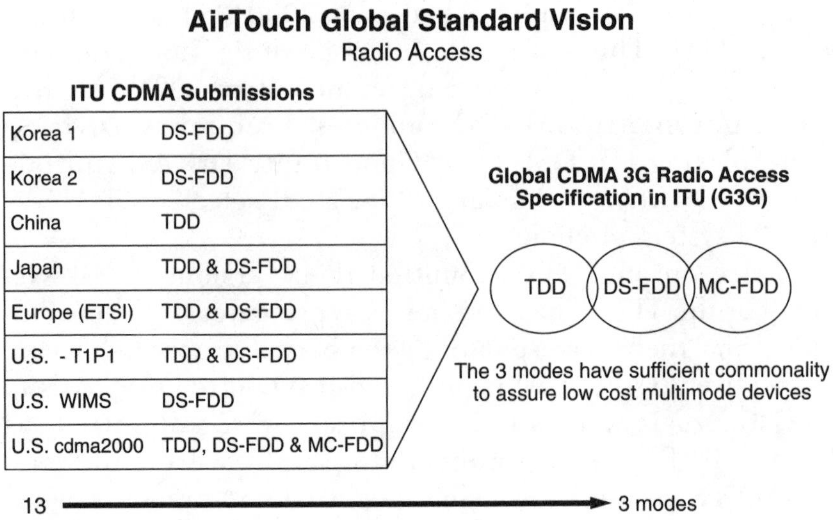

AirTouch Global Standard Vision
Radio Access

ITU CDMA Submissions

Korea 1	DS-FDD
Korea 2	DS-FDD
China	TDD
Japan	TDD & DS-FDD
Europe (ETSI)	TDD & DS-FDD
U.S. - T1P1	TDD & DS-FDD
U.S. WIMS	DS-FDD
U.S. cdma2000	TDD, DS-FDD & MC-FDD

**Global CDMA 3G Radio Access
Specification in ITU (G3G)**

TDD   DS-FDD   MC-FDD

The 3 modes have sufficient commonality
to assure low cost multimode devices

13 ──────────────────────► 3 modes

Figure 4.3.   ITU CDMA submissions.

harmonize from 13 modes to a few fundamental modes was tried. A new ad hoc organization made up of global operators, called Operator Harmonization Group (OHG), was formed in October 1998. The operators took the lead and converged 13 modes to 3 after many negotiations. The following five major meetings created this harmonization:

January 1999	First OHG meeting "harmonization framework agreed" in Beijing
March 1999	Second OHG meeting in San Francisco
April 1999	Third OHG meeting "Baseline Technical Agreement" in London
May 1999	Fourth OHG meeting in Tokyo
May 1999	Fifth OHG meeting "final technical agreement" in Toronto

On June 13, 1999, the 3G technical agreement from the OHG's harmonization effort was ratified. The three modes are shown in Fig. 4.4. The details are covered in Sec. 7.2.

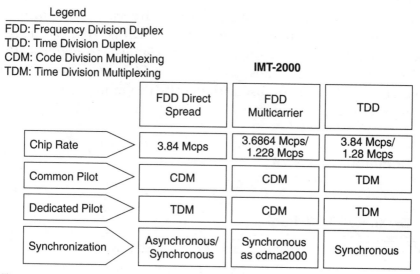

Figure 4.4. 3G harmonization agreement.

4.12 3G RADIO TECHNOLOGY OVERVIEW[7]

After the harmonization effort, four areas were reaching consensus.

4.12.1 COMPROMISED CHIP RATE

For CDMA-DS, the chip rate is 3.84 Mcps and for CDMA-MC the chip rate is 3.68 Mcps. It is easy to have filter implementation with two different chip rates, closely separated in every 5-MHz band. However, the 3.84 Mcps is not compatible with MC mode and cdmaOne.

In November 1999, the FDD multicarrier mode had a low chip rate version of 1.228 Mcps (1.25 MHz) and TDD had a low chip rate version of 1.28 Mcps (1.6 MHz), both of which had to be officially standardized.

4.12.2 PILOT STRUCTURE

In CDMA-DS, the common CDM pilot structure is the same as in cdma2000. However, some data channels can have dedicated TDM pilot signals that are assigned to different mobiles, as shown in Fig. 4.5. In CDMA-MC, the dedicated CDM pilots are auxiliary pilot signals associated with individual data channels, as shown in Fig. 4.6. These dedicated CDM pilots are used to identify the multiple beams in one sector for the smart antenna implementation.

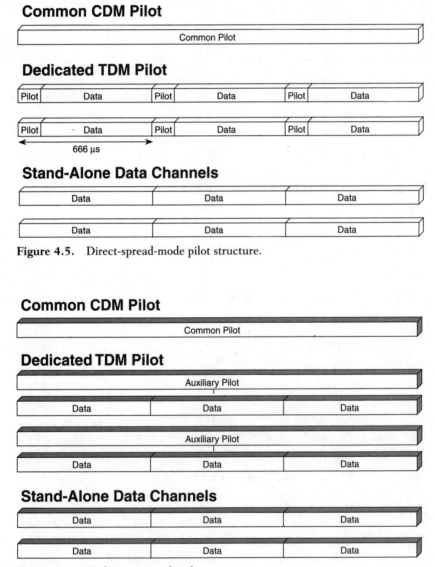

Figure 4.5.   Direct-spread-mode pilot structure.

Figure 4.6.   Multicarrier-mode pilot structure.

4.12.3   CELL SEARCH

MULTICARRIER (MC) MODE CELL SEARCH   Assume that cell A is the current cell. In cell A there is one common cell ID code, which we may call DL (downlink)) scrambling code or pilot code. A delay time with the same pilot code can be used to identify a different cell, as shown in Fig. 4.7.

DS MODE CELL SEARCH FOR SYNCHRONOUS OPERATION   Assume that cell A is the current cell. In cell A there are two kinds of synchronization channels, primary synchronization channel (PSC) and secondary synchronization channel (SSC) residing in every time slot. There are 15 time slots in each 10-ms window, as shown in Fig. 4.8. In each channel those 15 time slots use the cell ID code, which may be the DL scrambling code 0. After initial access, the timing relationship with neighboring cells is known. The same cell-search procedure can be applied to the MC mode.

DS MODE CELL SEARCH FOR ASYNCHRONOUS OPERATION   In this mode, there are three steps in cell search for initial access:

1. Find the slot timing from the PSC using matched filtering.

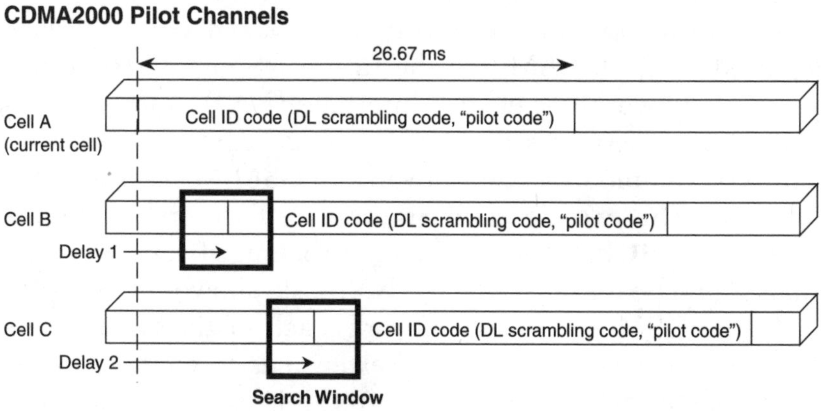

CDMA2000 Pilot Channels

Figure 4.7.   Multicarrier mode cell search.

WCDMA Synchronization Channels

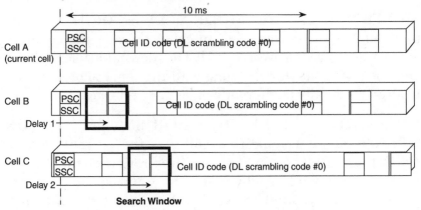

Figure 4.8. DS-mode cell search: synchronous operation.

2. Find the frame/code timing and code group from the SSC.
3. Decode the cell ID.

In the current cell A, the 15 time slots are spread in a 10-ms time interval, as shown in Fig. 4.9. This particular mode may be used in a closed-in area such as an underground shopping mall.

4.12.4   3G NETWORK CONNECTIONS

There are two major core networks, the ANSI-41-based core network deployed for both cdmaOne and the future MC-mode access networks and the GSM-MAP-based core network deployed for both GSM and the future DS-mode access networks. A network-to-network interface (NNI) connects DS-mode access networks to ANSI-41-based core networks and connects MC-mode access networks to GSM-MAP-based core networks. The two standard groups, 3G Partnership Project (3GPP) and 3GPP2, are developing the specification for the "hooks" and "extensions" for the NNI task, shown in Fig. 4.10.

In the GSM-MAP core network, each of the three layers (L1, L2, and L3) has hooks and extensions to connect to the corresponding three layers of the ANSI-41 core network. The

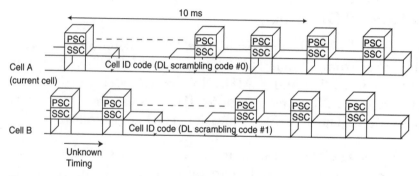

Figure 4.9.   DS-mode cell search: asynchronous operation.

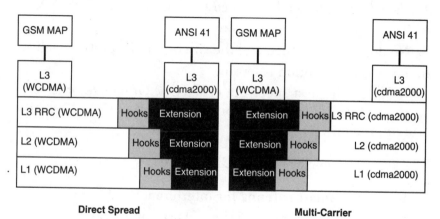

Direct Spread   **Multi-Carrier**

Figure 4.10.   Hooks and extensions.

connection from the ANSI-41 core network to the GSM-MAP network will have the same approach. The hook provides the specified functionality in the initial release of the standards for future extensions without major protocol changes. The extensions provide the additional function needed to meet the specified requirements when hooks are in place. A detailed description is in Sec. 7.2.

CDMA TDD access networks will connect to GSM-MAP core networks and will need hooks and extensions to interface to the ANSI-41 core network at each layer of the three layers (L1, L2, and L3).

4.12.5 TDD MODE

There are two ITU proposals, one from Acatel/Siemens and the other from China. The Chinese TD-SCDMA system has two versions, 1.6 and 5 MHz. In the 1.6-MHz version, the chip rate is 1.28 Mcps. In the 5-MHz version, the chip rate is 3.84 Mcps. The advantages of using TDD are as follows:

1. Any single spectrum band with a bandwidth equal to or greater than 1.6 MHz can be considered for the TDD application.
2. The TDD can dynamically handle asymmetrical traffic.
3. The research and development cost for the system is relatively low.
4. The handset cost can be lower.
5. No duplexer is needed and the size of the handset is smaller.
6. Power consumption is lower.
7. It has the highest spectrum efficiency. Of course, it is relatively hard to meet the power amplifier (PA) linearity requirement.

The key technologies in TD-SCDMA are

1. Use of the smart antenna for interference reduction.
2. Use of multiple time slots (i.e., TDMA and DS/CDMA).
3. Use of synchronous CDMA.
4. Multiple detection to enhance the detection scheme.
5. New interference cancellation scheme.
6. Baton handoff is used to take advantage of hard and soft handoff. It is suitable for TDD mode.

The main specification in TD-SCDMA is listed in Table 4.1.

4.13 CONCERNS OF 3G DEVELOPMENT

4.13.1 VENDOR'S CONCERNS

1. *Political situations.*   The United States government encourages innovation; therefore, IPR becomes a great issue for the radio access technologies, which delays the

Table 4.1. Main Specification in TD-SCDMA

Carrier separation	1.6 MHz
Chip rate	1.28 Mcps
Duplex adaptive TDD	5-ms interval
Multiplex	SDMA + CDMA + TDMA
Time slots number	10/7
Spreading factor	1/2/4/8/16
FR modulation	QPSK
Basic data rate	1.2/2.484.8/9.6 kbps up to 384 kbps
Voice	8 kbps
Maximum data rate	Up to 2 Mbps (asymmetrical)
Features	User positioning, handoff, and roaming

3G development. In Europe, the community put their collective effort into making a global wireless core network so successful with their GSM radio access. They would be reluctant to move to any different core network.

2. Too many different organizations, such as UMTS, 3GPP, 3GPP2, OHG, would slow down the 3G development.

3. *Spectrum issues.* The issued IMT2000 spectrum cannot be used in the United States. Therefore vendors would like the multiband spectrum solution to be adapted by the 3G. Also a few new spectrum bands for global 3G systems are being considered.

4. In developing a seamless system among three modes, the work on hooks and extensions is a great concern.

5. *Develop a common vocoder.* In the future GPRS (extended to packet switch from GSM system) will use the AMR vocoder. CdmaOne is using the EVRC vocoder. These two are quite different. Also, cdma2000 vendors are developing the SMV, another vocoder that can reduce the code rate even lower. Now it is desirable to have a common vocoder for global roaming purposes.

6. The development of an IP-based core network is beneficial to the operator, as stated in Sec. 8.7.

7. The cost, latency, and mature system timing issues of 3G development are not known by the operators. The vendors do not have any idea about them either. It is a very uncertain system we are facing.

4.13.2   OPERATOR'S CONCERNS

1. CDMA operators are at a crossroad. Here are a few questions that the cdmaOne operator may ask:

   a. Which of the three modes of 3G should be pursued?
   b. Should cdma2000 1x be deployed today?
   c. Will cdma2000 be a technical winner for 3G?
   d. Will Asia-Pacific be a large market for cdmaOne or cdma2000 1x in the near future?
   e. What will be the advantage of the cost reduction in 3G transition?
   f. Will the customers be pleased with cdma2000 (CDMA-MC)?

2. GSM/TDMA operators are at a crossroad as well:

   a. Will GPRS or EDGE be the right interim step toward 3G?
   b. Will CDMA-DS be a low-risk developed mode?
   c. How can we make the transition more economical?
   d. Can cdma2000 be considered?

4.14   THE FUTURE OF WIRELESS COMMUNICATIONS BEYOND 3G

Wireless communications and the Internet are growing rapidly. Today there are 500 million mobile users and 200 million Internet users worldwide. It is inevitable that wireless communications and the Internet will combine to meet the requirements for the future. The Internet will move toward a wireless Internet. The network will have a wireless IP core. In the future, any information available worldwide will be acquired from one device at any time, anywhere, as shown in Fig. 4.11.

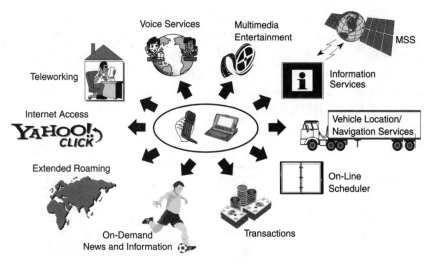

Figure 4.11. Any information to any device, at anytime.

4.14.1 THE MULTIDIMENSIONAL MOBILE COMMUNICATION SYSTEM, A BROAD CATEGORY OF SYSTEMS

1. Multi-information media system that includes telecommunication, computer communication, and entertainment communication

2. Multitransmission media system that includes voice, data, video, and image

3. Multilayered network that includes terrestrial mobile, satellite mobile, and air-to-ground

The system has to be in a gigahertz frequency range for high capacity and in a broadband for high data rate.

4.14.2 RADIO ACCESS

The propagation range under high frequency and a broadband system requirement is limited. Therefore we need to have at least a 50- or 100-m solution that can use millimeter waves or infrared links. To deliver a broadband message, a hybrid network will be needed to and from the wire-line to the wireless link, which is the last 100-m (or less) link. Also, the LOS links can be provided by the LEO satellites, geosynchronous high-altitude aircraft platforms for microwave, or millimeter wave communications. From these approaches we can provide not only high speed but also broad bandwidth (large capacity) services.

4.14.3 KNOWLEDGE-BASED NETWORK

As we head into the information age, the knowledge-based network becomes increasingly important. Knowledge is stored in databanks. This data can be managed in four different operations: production, storage, transfer, and application. In the transfer operation, the use of compression, caching, and prefetching are among the new techniques available to enhance a knowledge-based network.

4.15   THE DREAM OF DEVELOPING 4G

Vendors have been the driving force behind developing 3G. Now the operators feel the situation is

$$\text{Vendors} \xrightarrow{\text{tell}} \text{service providers (operators)} \xrightarrow{\text{tell}} \text{users.}$$

In the 3G standard groups, 3GPP develops the FDD-DS system and 3GPP2 develops the FDD-MC systems. Now the operators are starting to have many concerns and fears about how the vendors are developing and implementing the 3G systems. Will the customers like the 3G system? Can operators control the cost of the 3G system? If the operators do not like this compromised, multimode system, what is the next step? The answer is to hope that the 4G system can be an ideal system.

The ideal 4G system will be a single mode system. It should be driven by public interest. The process should be

$$\text{Users} \xrightarrow{\text{tell}} \text{service providers (operators)} \xrightarrow{\text{tell}} \text{vendors}$$

In this way, users will be happy, and this will create more business for service providers and vendors. Of course, vendors will need to comprehend this new approach, and it will be the operator's job to creatively lead the way.

4.16 REFERENCES

1. W. C. Y. Lee, "Dual-Mode Capability in Cellular Communications," *Communication,* Nov. 1987.

2. W. C. Y. Lee, "Cellular Operators Feel the Squeeze," *Telephony,* May 30, 1988, pp. 22–23.

3. Bell Labs, "High Capacity Mobile Telecommunication System Developmental System Reports," No. 1–No. 8 published every 3 months from March 1977 to March 1979, submitted to FCC.

4. Cellular Systems, IS-54 "Dual-Mode Mobile Station–Base Station Compatibility Standard," EIA, Engineering Dept., December 1989.

5. Cellular System, IS-136 "800 MHz TDMA Cellular—Radio Interference—Mobile Station—Base Station Compatibility," TIA/EIA, December 1994.

6. OHG, "Harmonization Framework Agreement for 3G" Ottawa, Canada, June 3, 1999.

7. 3GPP's 3G Specification, ITU IMT-2000 Workshop, Toronto, September 10–11, 1997.

LEARN FROM THE PAST

5.1   DUOPOLY COMPETITION[1]

The spectrum is a limited natural resource and today is a precious commodity. From a technical standpoint, to have spectrum efficiency, the entire allocated spectrum should be used to conduct only one service. However, for the fairness of competition, two or more operators are needed to create a balanced situation. Allowing too many competitors will leave no room for competition, as each one would only operate a small band.

Based on the above, allocating a 20- to 30-MHz band per operator in cellular or PCS is a good choice. In the past, four licensees, each with only a 1-MHz band in a UK CT-2 system, could not run their wireless phone services at a profit. This means that too many competitors and too little allocated spectrum could kill the service entirely.

Also, in duopoly competition, two companies are always trying to compete to gain market share. However, there is the unique interference phenomenon in the cellular system. As more customers are served in a system, the interference becomes greater and the voice quality of the system declines. Therefore, a new customer wants to go to a system that has a low market share, because its voice quality is better. As a result, the market shares of two systems in any market are very close regardless of their effort in marketing.

In 1987, Los Angeles Cellular Telephone Company (LACTC) started its service in Los Angeles and advertised that its system served digital voice. In actuality it used the Ericsson digital switch, but the radio voice was still analog. The customers did not know that the voice quality was good because there was only a small number of customers at the beginning of its commercial service, which made the call-traffic interference level low in its voice channel.

5.2   IMPACT OF MFJ[2]

In 1980 Judge Greene from DOJ was appointed to divest AT&T based on the antitrust law. AT&T had to give up its 26 regional Bell operational companies. These 26 companies were consolidated into seven regional Bell companies, called RBOCs. They were Pacific Telesis, Nynex, Bell Atlantic, Ameritech, US West, Bell South, and South Western Bell. The DOJ brought the first antitrust suit against AT&T before some of us were born. Here's an overview of key dates:

MFJ TIMELINE

January 14, 1949	DOJ files antitrust suit against AT&T.
January 24, 1956	Final Judgment entered.
November 20, 1982	Modification of Final Judgment (MFJ) entered.
January 1, 1984	Divestiture of RBOCs.
February 2, 1987	DOJ files first triennial report on the MFJ.
September 10, 1987	Judge Greene relaxes information services restrictions.
December 3, 1987	Judge Greene clarifies manufacturing restriction; design prohibited.

The agreement between AT&T and Judge Harold H. Greene was known as MFJ and it was signed on August 24, 1982. In MFJ, RBOCs had three restrictions:

1. Provide interexchange telecommunication services or information services
2. Manufacture or provide telecommunications products or customer premises equipment (except for provision of customer premises equipment for emergency services)
3. Provide any other product or service, except exchange telecommunications and exchange access service, that is not a natural monopoly service actually regulated by tariff

Further, the "no manufacturing business" portion of MFJ stated that

1. RBOC cannot instruct the vendors about how to design or develop products.
2. Any manufacturing products can only be produced outside

the United States and can only be sold outside the United States.

3. Outside the United States does not include Mexico and Canada. If the RBOCs violated the MFJ, the court would consider it a criminal act.

The advantages of MFJ was that the RBOCs had no manufacturing capability so the small wireless communication manufacturing companies had a chance to get into the market. The disadvantages of MFJ included

1. RBOCs had no choice but to use the vendor's equipment.
2. RBOCs' engineers ran the market operation, understood the problems, and were trained in finding the solutions but could not pass those solutions along.

The following are stories about the impact of MFJ:

1. *In-building cellular system invented in 1986.*   On July 23, 1986, W. C. Y. Lee asked consultant R. A. Isberg to quote the cost of building a prototype of Lee's idea, which he drew on a napkin at the San Francisco airport cafeteria. Later Isberg redrew Lee's system diagram as shown in Fig. 5.1 and Exhibit 5.A. Pacific Telesis MFJ attorney found the drawing after Lee received it and feared that the drawing might violate MFJ. As an officer of the Pactel Cellular Company, Lee was advised to sign a letter and send it to Isberg. In the letter, Lee experienced the willingness to give up his patent right so no MFJ issue would be involved. The inbuilding system was developed in 1987, but at that time nobody from a RBOC dared to commercialize an invented system in view of MFJ. This consultant-made system sat in Lee's laboratory. In 1988, Sam Ginn, Chairman of Pacific Telesis, remembered Lee's early presentation on the in-building system and asked that it be installed in the Pacific Telesis building. It took 3 weeks of preparation. On September 9, 1988, the system was operated under the Experimental License issued to Cellular One, San Francisco. The Low-Power Cell Enhancer made by Isberg is shown in Fig. 5.2. The Yagi Antenna was

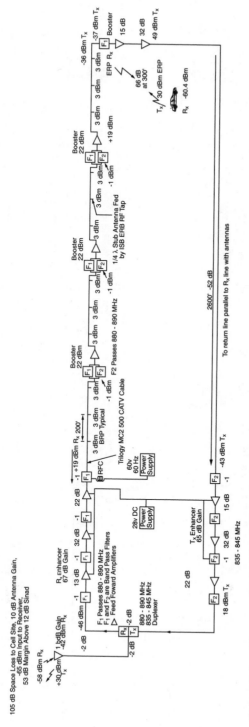

Figure 5.1. Diagram of inbuilding cellular system.

123

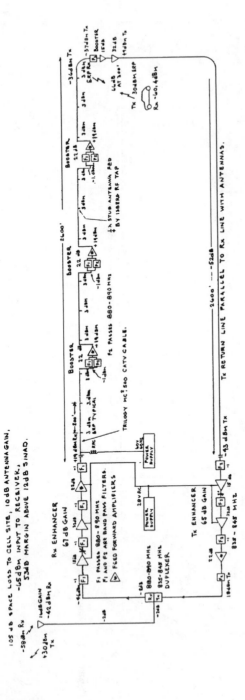

Exhibit 5.A. A first inbuilding communication system.

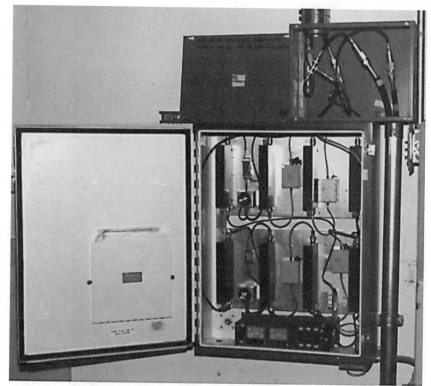

Figure 5.2.   The low-power cell enhancer.

mounted on the roof terrace and used for transmitting to and receiving from Cellular One's cell site, as shown in Fig. 5.3. The antenna mounted in a garage is shown in Fig. 5.4.

Motorists driving into underground garages could continue cellular calls without call drops. All the executives were very happy when they could continue their calls in their limousines while entering underground garages.

2. *Prototype of low-cost enhancer in 1988.*   Lee, at Pactel Cellular, tried to design a low-cost, small-size enhancer that could be easily mounted on any utility pole. The enhancer, also called a repeater, enhances the signal after it is received. In 1985, the commercial enhancers were as big as the refrigerators. Then in 1988 the size was reduced, but was still too big. Lee made 10 miniature ones as shown in Fig. 5.5 and planned to deploy them in Pactel's own system. The prototype unit was

Figure 5.3. The Yagi antenna mounted on a roof terrace.

mounted on the utility pole in the Luguna Canyon at Luguna
Hill, Orange County. The result was very good. But Pactel's
MFJ attorney advised Lee not to put it in service because Pactel
was not allowed to manufacture any products even for internal
use. Later, in 1991, Pactel had made an agreement to have
Avantek manufacture this low-cost enhancer. Unfortunately,
Avantek was bought by HP and the deal was canceled.

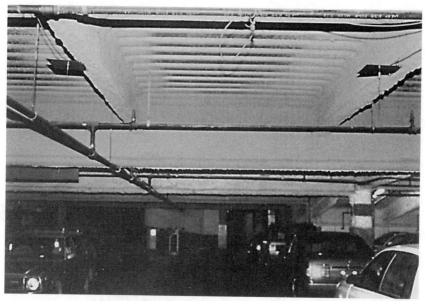

Figure 5.4.   Another antenna mounted in a garage.

3. *The PCN antennas made for UK Microtel.*   In 1989 Pactel
sent Lee with the technical team to England to apply for a
Personal Communication Networking (PCN) license, which
they won. The new consortium company was named
Microtel. The PCN frequency was 1800 MHz. At that time,
no commercial base stations or mobile stations were on the
market. Microtel was in a hurry to acquire one 1800-MHz
base-station antenna from Canada, which cost about $5000.
Lee heard about this and immediately asked his engineers to
make ten antennas; two for the base stations and eight for the
mobile station in 3 weeks. The cost was about $30 (shown in
Fig. 5.5). Lee thought that products made in the United
States but used outside the United States would not violate
MFJ. Lee planned to charge Microtel $500 each for the
antennas, and the price was agreed on. However, Pactel
MFJ's attorney advised Lee that according to MFJ, the 10
antennas could not be sold even outside the United States.
The alternative was to loan the 10 homemade antennas to
Microtel for 90 days. In this way, Lee and his team did not
get any monetary reward.

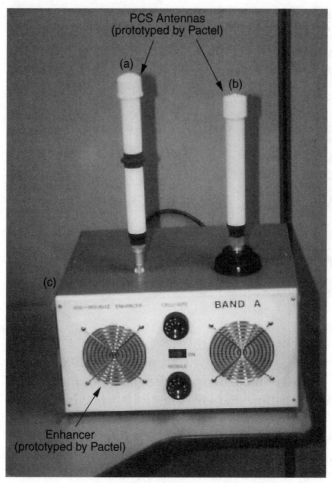

Figure 5.5. A prototype of a low-cost, small-size enhancer.

4. *Planning setup and manufacturing in Tijuana, Mexico in 1989.*   Because Pactel had a cellular market in San Diego, it was easy to set up manufacturing in Tijuana, which is 40 miles away from San Diego. Manufacturing and selling products outside the United States should not violate the MFJ. The statement says "outside the U.S.," but as the Pactel MFJ attorney explained, that did not include Mexico and Canada. Lee at this time totally gave up.

5.3   HISTORY OF WHY NO "CALLING PARTY PAY" FEATURE

Many engineers and executives have said that the advancement of the European cellular systems were benefitted by the calling party pay feature and in the United States cellular system it was not. Today the United States' cellular systems can only have "mobile party pay" for the airtime on both directions. This means the mobile party pays the airtime for both calling (outgoing) and called (incoming or receiving) parties. Because of this arrangement, cellular subscribers are not willing to give their numbers to anyone unless they can justify paying for the call. As a result, the promotion of minutes of use in this country cannot be as high as in Europe. Does the United States have the technology or not?

In 1976, AT&T thought cellular systems would be their line of business because they invented it. They planned to deploy the cellular system in each of the 21 RBOCs. Because each regional cellular business would be under one regional company, the billing functions of incoming and outgoing calls could be located at Class 5 central office (C.O.), which is the wire-line central office. Unexpectedly, in 1982 DOJ ordered the divestiture of AT&T and ordered all RBOCs to separate the cellular business from the landline business. Because the mobile telephone switching office (MTSO) is not a Class 5 C.O., it does not have a record of the telephone number of incoming calls. Thus the MTSO added a billing system called automatic management accounting (AMA) to bill airtime on both outgoing and incoming calls to the mobile phone number because it was unable to obtain the record from Class 5 C.O. to bill airtime on incoming calls. After divestiture, the wire-line companies were reluctant to let the MTSO be a Class 5 C.O. unless the government mandated it. The "calling party pay" feature that could not be implemented was due to a business reason not a technical one.

5.4   RESELLERS

The reseller is one who does not have his or her own operating system but wants to be a service provider. In 1984, many non-wire-line (not a telephone company) companies obtained their licenses but the systems were not ready to operate. They obtained mobile phone numbers from FCC and asked the wire-line (telephone company) companies who ran the cellular systems to serve their customers. For instance, the LACTC was the reseller that resold Pactel's service in 1985 and 1986. It had a chunk of mobile numbers on Pactel's switch. It paid a discount fee to Pactel and billed its customers directly with its own rate. The AMPS phone has a feature, Prefer A and Prefer B, meaning prefer using Band A (non-wire-line band) or prefer using Band B (wire-line band). LACTC's phones were always set at Prefer A. When LACTC users called on Band A and there was not a Band A system, the call immediately went to the Band B system.

This was the FCC's way of maintaining fair competition because the Band A operators' systems were deployed in the markets late. Therefore, LACTC did not need to push its cellular system into operation sooner because of reseller status. The growing number of customers burdened Pactel switches. Operating a new inexperienced system in the Los Angeles market and trying to keep up with a rapidly growing customer base was very difficult. Many unexpected problems developed from capacity issues and frequency management to handoffs and other issues. Trying to maintain voice quality while customers were increasing was a big challenge. Every time voice quality improved, an increase of traffic would degrade it. In addition, the reseller's customers made things more difficult for the system operator. Pactel kept pushing LACTC to open its own system. There were several delays in their schedules. If the FCC did not have a reseller policy, the situation would have been different. LACTC would have worked much harder to operate its system earlier to take care of its customers, and Pactel would have wished that LACTC could not delay its system operation.

When LACTC opened its system during the second half of 1986, it shifted its customers to its switch. Because they had only about one-fifth the customers that Pactel did, LACTC's system experienced fewer interference problems due to a lighter load in the beginning of its service. The company claimed that they had a better-quality system because the voice quality was superior. Customers did not know that the light traffic load improved voice quality and LACTC gained many customers. But once the customer volume started building up, the voice quality started to degrade due to the interference generated by the traffic load.

5.5   PACTEL MOVES TO BAND A

Before 1985, the FCC made the rule that nontelephone companies (called nonwire) could only operate on Band A, and telephone companies could only operate on Band B. After 1985, Pactel began to challenge this FCC rule and purchased Communication Industry (CI), which owned the Band A markets such as San Diego, San Francisco, Atlanta, and a minor partnership in other Band A markets. CI also owned two manufacturers, BBI, a paging manufacturing company, and DB Products, a radio manufacturing company. Pactel was a subsidiary company of Pacific Telesis, one of seven Baby Bells. The first hurdle for Pactel was to get approval to operate on Band A from Judge Greene, who was handling the divestiture of AT&T. The conditional approval was

1. Pactel could not own two licenses in the same market.
2. Pactel could not run a manufacturing business, based on the MFJ restriction.

Pactel sold the San Diego market to the US West and got rid of BBI and DB products in 1986.

In 1986, the FCC granted a waiver to Pactel to operate in Band A markets. The value of each market in 1985 was hard to determine. Then a term called $/POP was used, which referred to the purchase price divided by the population of the market. Pactel bought CI at a price of $32/POP. In 1985, the cellular industry said the price was too high. Later, in 1986, the price of CI deal was considered to be very low because other Band A markets were sold at a higher price.

After Pactel purchased CI and operated in Band A markets, all other RBOCs followed suit and ran Band A markets. Since the "non-wire-line band," Band A, could be operated by the wire-line companies who supposedly could only operate on the "wire-line band," Band B, the terms *wire-line band* and *non-wire-line band* have been abandoned and have no meaning today.

5.6 SINGLE-STANDARD SYSTEM VERSUS MULTISTANDARD SYSTEMS IN CELLULAR

At the beginning, the first-generation cellular system in North America was a single-standard system called AMPS. But in Europe, each country had its own mobile radio system. There was no single-standard system in the continent of Europe. Then Europe developed the single-standard GSM for the second-generation digital systems, but the United States turned out to have four different systems, analog, TDMA, CDMA, and DCS1900. If there is only one standard, a concentrated effort can continuously make this standard system better. The GSM system today, although it is not a high-capacity system, has already penetrated the world, as the U.S. analog system did during its beginnings. With the four different systems and two different bands, 800 MHz and 1900 MHz, the roaming issue becomes very critical. All the systems in the United States need to use the analog channel as the foreign system roaming channel. In Europe, they could withdraw the analog systems. The United States cannot, especially for the E911 service. We have to rely on the only nationwide standard, the analog system. This analog system will survive until the 3G is fully deployed. In the meantime, the U.S. second-generation systems cannot compete with GSM's system. Besides, each of the four systems has a different radio waveform and often can interfere with each other. A way to reduce the interference among them would cost a lot and may never be found.

5.7 SPECTRUM SHARING

The policy of sharing spectrum increases spectrum efficiency. Sharing the spectrum for two or more services that are within one operator's control is the best policy. Sharing the spectrum with two different operators for two different services needs great caution. For example, a possible harmful interference to the cellular system from an air-to-ground service[3] (ATGS), as shown in Fig. 5.6, results because ATGS is trying to share the same 800-MHz cellular spectrum. The ATGS's horizontal polarized signal in the reverse link (aircraft-to-groundstation) can strongly leak into the vertical polarized

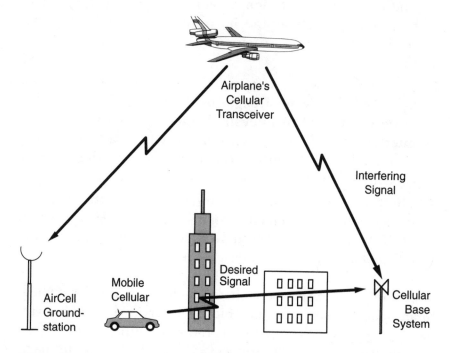

Figure 5.6. Harmful interference from the airplane system to the cellular base station.

antenna at the cellular base station, as shown in the measured data.[3] Also the ATGS power control is unable to effectively reduce harmful interference from the cellular system. As seen in Fig. 5.6, when an aircraft flies closer to the cellular base station but away from ATGS groundstation, the ATGS power control increases the aircraft's transmit power due to its weak reception at its groundstation. This increase of power creates a harmful interference to the cellular base stations.

Besides, the 800-MHz spectrum allocated for cellular has demonstrated its capacity efficiency as compared to any other service in other spectrum bands. The future cellular services are especially prone to harmful interference as follows:

1. The E911 service will be provided by cellular operators in 2001, mandated by the FCC. This system cannot afford any interference. It is a matter of life and death.

2. The cellular base-station antennas are going to use two polarization, vertical and horizontal, instead of only the vertical polarization to improve the signal reception from the mobile.

3. The digital CDMA system, whose traffic signal is one-twentieth the signal strength of the analog signal, will be more vulnerable to undesired interference. Any foreign interference will reduce system capacity in a CDMA system. Adding 3-dB interference will reduce the CDMA capacity by one half.

4. The further improved spectrum-efficiency system, the 3G system, will be implemented in cellular 800-MHz spectrum in 2001. The new spectrum-efficiency system cannot be affected by any foreign interference.

Therefore, the spectrum-sharing policy can only be applied to an allocated spectrum in which one service has no growth in usage, or the polarization separation is sufficient for interference isolation for two services. Of course, if one primary service needs continuous improvement and it also uses the new technology with the two polarizations for achieving the spectrum efficiency within its own spectrum, then the FCC cannot stop allowing the new polarization technology applied by the primary service in order to carry out the sharing spectrum policy with the secondary service.

5.8   WHY NO DIVERSITY RECEIVER AT THE MOBILE STATION

In 1970, Bell Labs designed space-diversity schemes and put one at the base station and the other at the mobile station. At the base station, the antenna spacing was eight wavelengths (or 9 ft at 850 MHz), which is used today. At the mobile station, the antenna spacing is a half a wavelength (or 6 in at 850 MHz). Two antennas, separated by 6 in, can be mounted on the roof of a mobile station without any difficulty. Then the system designer can decide to use different combine techniques for the diversity signals. There are two families of combining techniques used after receiving two signals from two antennas:

1. Using two receivers, the techniques are maximum (signal-to-noise) ratio combining, equal (cophase) gain combining, and selective combining.
2. Using one receiver, the technique is switched combining, the switching being based on a predetermined threshold level.

Using two receivers for combining the received signals results in a better performance, but the cost of having two receivers is high. Using one receiver for combining provides a low-cost solution, but the performance is not desirable. Often the performance when using one-receiver diversity exceeds that of using the one-receiver nondiversity when the signal is acceptable.

However, the performance of one-receiver diversity is noticeably worse than that with nondiversity when the signal is weak. The diversity is supposed to help the weak signal, but it does not. At the beginning, F. E. Johnson, Motorola, and NEC made mobile diversity receivers. Because the performance of one-receiver diversity was not good at the weak signal spot, nobody liked them. The mobile diversity receivers were discontinued.

The selection of diversity receiver based on merely a low cost taught us a lesson. When designing a new system or appliance, do not consider the cost if the risk of low performance may occur. A high cost but better performance, just like AMPS, always wins the business in the long run. The cost of a mobile unit was more than $3000 in 1984. The cost came down because of purchase volumes and technology advancement. In 1999, the handset cost is around $150.

5.9   ROOF-TOP ANTENNA ON VEHICLE

At the beginning of the cellular system in 1983 and 1984, system operators recommended that mobile antennas be mounted on the middle of the roof of the vehicle for good reception because the roof is made of metal, which can be used as a good ground plane. Mounting antennas on the roof would allow antenna patterns to be uniformly radiated 360° horizontally. But to mount the roof-top antenna, a hole needed to be drilled into the car roof to maintain a wire connection between the antenna and the mobile receiver. Most customers didn't want their car roof drilled, so when glass-mounted antennas came on the market, customers liked them. The reception of glass-mounted antennas, however, was 3dB weaker than the roof-top-mounted variety because the electrical path was using the inductance to virtually connect on the two sides of the back window—one side with an antenna and the other side with the receiver. The customers were willing to suffer some quality degradation in exchange for not having a hole in the car roof.

Gradually all roof-top antennas were replaced by the glass-mounted antenna based on customer requests. Roof-top antennas were no longer on the market. And customers soon forgot the 3-dB loss due to the glass-mounted antenna and were complaining about the service performance so often that system engineers tried to find a way to make up the 3-dB loss.

5.10   NO GOOD DATA MODEM FOR AMPS

In 1984, wire-line data modem manufacturing companies were trying to design modems for the cellular application. They thought the only difference in design between the wire-line modem and cellular modem was the handoff feature in the cellular modem. The transmission rate was 300 bps. A portion of data during the short burst of 100 ms throughout the handoff process was lost and needed retransmitting. However, the quality of the cellular modem was very poor. The reason for this was as follows:

1. The transmission environment for the mobile radio transmission is different from the land-line transmission. The mobile radio environment generated the burst error in the signal due to the multipath fading distorting the signal, whereas the land-line environment generated the random error in the signal due to the gaussian noise in the environment. When the vehicle traveled faster, the length of the burst became smaller. When the vehicle traveled more slowly, the length of the burst became longer.

When data is sent in this medium, the existing land-line data transmission protocol needs to be reconstructed to cope with this medium so that the burst errors can be converted to the random errors, for example, by applying interleaving structure into the protocol.

2. The AMPS system for the mobile radio transmission is different from the fixed wireless transmission, which does not experience the fading phenomenon. The AMPS is an FM system designed in a multipath fading environment primarily for voice. The AMPS is not designed for data.

The FM transceiver has a device called preemphasis at the transmitter and one called the deemphasis at the receiver. The purpose is to send the voice (300 to 3000 Hz) so that it matches the same power distribution curve shape over the frequency spectrum as the noise distribution curve but is 30 dB or more higher. The noise distribution curve in the FM system looks like a parabola curve. As a result, the voice signal-to-noise-ratio (S/N) then keeps a constant over the frequency spectrum, and the high end of voice frequency will not be hurt by the noise while propagating through the medium. In addition, a syllabic compounder is used to compress the voice signal while transmitting and expand the voice signal while receiving. It can confine the voice energy within the channel bandwidth. The greatest advantage of using the compounder is that the audio output of the receiving end is much quieter due to the expander pushing the noise level down during the voice pause period. The preemphasis/deemphasis devices and compounders improve the voice quality but distort the data stream if the protection of data stream is not thought through. This was why the cellular data modem did not function as well as expected.

For these two reasons, the data modem without preamble or an interleaving scheme before sending would hardly be operable in the AMPS. This is why no successful modems were used in AMPS systems.

5.11   WHY NO OPEN INTERFACE STANDARDS

From 1983 to 1998 in the U.S. cellular industry, except CAI, no other open interface standard has been developed between any system elements; for example, there is no interface standard between BTS to BSC, BSC to MSC, or MSC to MSC. We have seen from GSM system development discussed in Section 4.5 that the A interface allows a system operator to use one vendor's BTS and another vendor's BSC or MSC. Competition among vendors in Europe brings the cost down. The size of equipment becomes smaller, and the quality becomes better. In the United States, however, radio modules are still very large, taking a lot of space, and the cost of equipment is still very high because there are no open interface standards and no competition. Once a selected vendor's system is developed, the system operator cannot replace any system elements with those from another vendor. The system operator is forced to stay with a particular vendor's system unless the operator decides to replace the entire system, which is economically infeasible. Therefore the operator has to pay a high price for any improvement on the system because no other vendors can modify this particular vendor's equipment. Operators are angry that handsets, because of competition due to the CAI standard, have been reduced to a business-card size, but the base-station radio still remains the same size as 5 or 6 years ago.

The U.S. cellular industry actually followed AT&T's past culture. AT&T set up its own system specification and sold its system as a whole. It never needed interface standards. In 1983, all AT&T customers were the RBOCs. They were from AT&T's family and trusted AT&T's products. The open interface standards were not brought up as an issue. Later many vendors, such as Motorola, Ericsson, Northern Telecom, and NEC, were in the cellular infrastructure business. The operators could not take advantage of vendor's competition because the situation of no open interface standards blocked the opportunity.

5.12 MILLIMETER WAVE AND OPTICAL WAVE LINKS

In 1972, the New York Telephony Company was facing call congestions in the network and needed to add more cables to the network. But digging in the city to lay the cables was very costly. Then they got the idea of using radio links instead to provide cable links; the radio frequency had to be in millimeter or optical waves. The former one would be attenuated heavily by rainfall and the latter would be attenuated heavily by fog. Then they developed the concept of using both links, millimeter and optical wave, side-by-side, simultaneously transmitting and receiving and providing reliable links against rainfall and fog.

The optical link was set up by B. G. King,[4] and the mil-

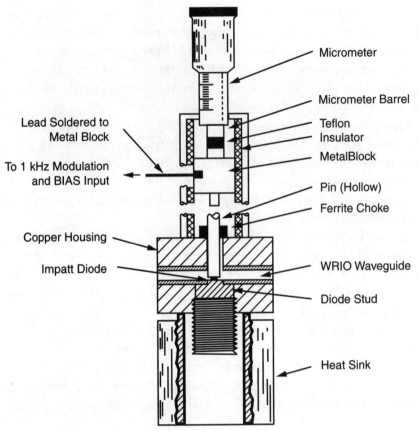

Figure 5.7.   100-GHz oscillator with 1-kHz modulation.

limeter link was set up by W. C. Y. Lee.[5,6] Lee's apparatus was built using the Bell Labs newly invented Inpatt diodes[7–9] as the 100-GHz (3-mm) source. This was the Inpatt diode's first application. The 1-kHz modulator modulated the source as an on/off switch. The oscillator shown in Fig. 5.7 was designed by Lee. The modulated signal was generated through the WR10 waveguide.

The 2.5-ft parabola 100-GHz (3-mm) antenna provided 60-dB gain but it cost $2000 every 6 months to smooth the parabola surface due to the dust. Any granularity (unsmoothed) defocused the antenna beam and reduced the gain. The antenna pattern of 0.3° is shown in Fig. 5.8. Then the apparatus shown in Fig. 5.9 was built. The phase reference between the transmit and receiving ends was linked by the telephone leased line. At the beginning the apparatus was tested at Holmdel, New Jersey. A record made on October 30, 1973, is shown in Fig. 5.10. Each dot of data represented an average of 2 minutes. The signal was fairly stable during the 8 hours of data except for 40 minutes between 1:30 PM to 2:00 PM, when there was a shower and it needed the optical link's help. The two pairs of links, one for the millimeter wave and one for the optical wave, were installed with one end on the eighty-eighth floor of the Empire State Building (see Fig. 5.11) and the other on the fifty-seventh floor of the Pan Am building, facing each other. The link distance was half a mile in New York City (Fig. 5.12). The two links worked successfully.[6,10] A letter from Bell Lab to the FCC (see Exhibit 5.B) informed the FCC that the millimeter wave link would commence on December 4, 1973, in New York City for a period of 1 year. In 1973, frequencies above 40 GHz did not need a license from the FCC. However, Bell Labs had to argue with the supervisor of the Empire State Building, who believed that the 3-mm wave would cause harmful radiation to the building. Bell Labs asked a group of scientists to convince the supervisor but they failed. Finally, Bell Labs gave up this project. The work has been described in a book by Lee.[11]

5.13   RAIN RATE STATISTICAL MODEL FOR THE U.S. REGION

When using the frequency above 10 GHz, the signal attenuation from rain rate becomes very severe because of water drop absorption. All the U.S. Weather Bureau data were rain accumulation data, not the rain rate data; therefore, in the beginning of 1974, L. R. Lowry of Bell Labs used a bucket to catch raindrops. As soon as the bucket was full, he dumped out the water and set it down again to catch more rain. A paper recorder recorded the dumping rate of the bucket. Ten buckets were distributed in the Holmdel, N.J. area. Every 2 weeks, the record paper had to be changed. In May 1, 1974, a seminar was held to develop a rain

FCC Radio Licenses--Notification to
Commission of Millimeter Wave Propagation
Experiment at Empire State Building -
File 36634-2

NOV 2 9 1973

Federal Communications Commission
Washington, D. C.   20554

Attn: Mr. Vincent J. Mullins, Secretary

Gentlemen:

This is to advise that we plan experimental use of
Experimental (Research) radio license KF2XBY (File
No. 3735-ER-R-71) in New York City, commencing on and
after December 4, 1973 and continuing thereafter for a
period of about one year. We propose to evaluate the
effects on propagation attributable to weather at millimeter wave frequencies. For this purpose, we propose to
radiate up to 6 mw of 100 Gc/s CW energy from a 30"
parabolic dish with Cassegrain feed. The 3 dB beamwidth
of this array is approximately 0.3°. In operation, this
system would be installed to provide line-of-sight transmission between the 88th floor of the Empire State Building
and the 58th floor of the Pan American Building.

We know of no interferences that this operation would cause
and are satisfied therefor that we should neither produce
nor sustain interferences except from the atmospherics, the
properties of which we plan to evaluate.

By copy of this letter we are notifying the cognizant
District Engineer of our experimental plans. Should any
additional information in connection with this project be
required, please advise us.

Very truly yours,
ORIGINAL SIGNED BY

HO-4362-JHC-ahj   N. Levine for
                                   E. F. O'Neill   11/29/73
Copy to   Executive Director
Engineer in Charge   Toll Transmission Division
2nd Radio District
Federal Communications Commission Copy to
748 Federal Building   Messrs. T.C. Cross   AT&T
641 Washington Street   S.D. Hathaway
New York, New York 10014   W.C.Y. Lee
                                           N. Levine
          DATE FILE COPY   M. J. Pagones
       BELL TELEPHONE LABORATORIES   D. O. Reudink
             INCORPORATED   L. C. Tillotson

Exhibit 5.B. Notification to the Commission of millimeter wave propagation
experiment.

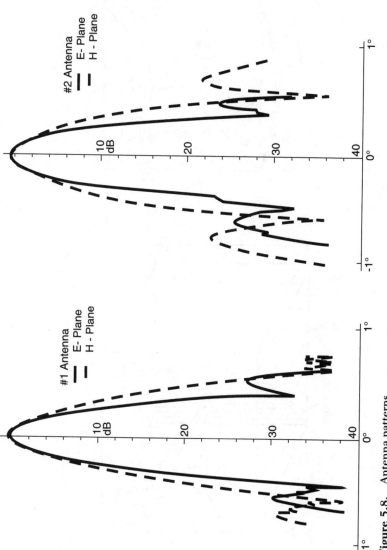

Figure 5.8. Antenna patterns.

143

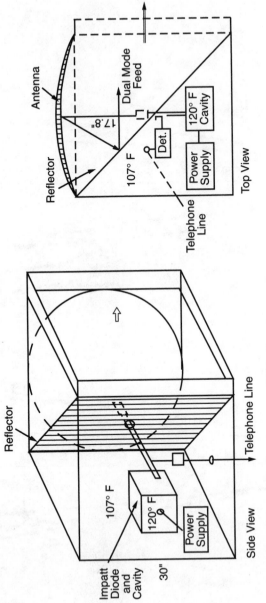

Figure 5.9. An overall setup on transmitting site.

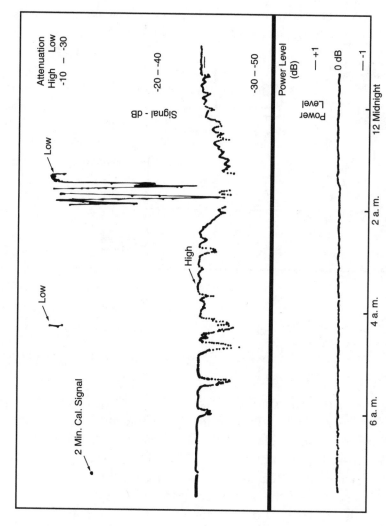

Figure 5.10. Recording data on Oct. 30, 1973 at Holmdel, N. J.

145

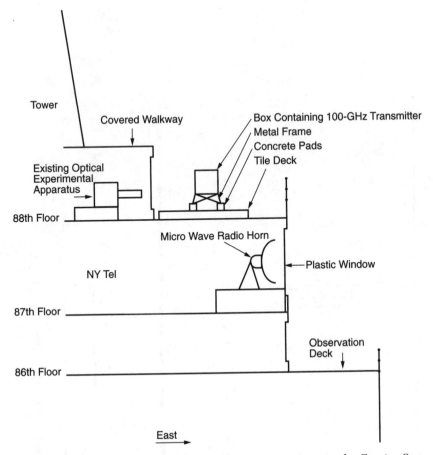

Figure 5.11. Location of the millimeter-wave transmitter in the Empire State Building.

rate gauge. After the seminar, Lee felt that this project was too costly and would take too long to obtain the meaningful statistics in a limited area. He went to Bell Labs library and took the U.S. Weather Bureau data from 1950 to 1973 for the rain accumulation each year. Because they were in 5-minute increments, Lee easily converted the accumulative data into rain rate data.

Also, the 23 years of data formed a statistical curve. The data was collected at 257 locations, mostly at the airport of cities across the entire United States. With all the data, a statistical model of rain rate in the United States could be determined. Lee wrote a memo[12] on May 10, 1974, after he checked with W. T. Barnett and his group at Bell Labs and was assured that this

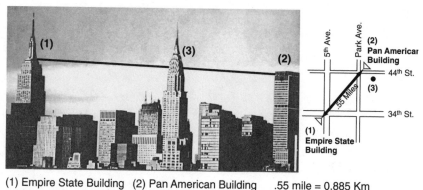

(1) Empire State Building   (2) Pan American Building   .55 mile = 0.885 Km
    Latitude = 40° 44' 54"   Latitude = 40° 45' 11"
    Longitude = 78° 59' 10" Longitude = 73° 58' 35"
    88th (1103') Floor   58th (846') Floor

Figure 5.12. Legend to come.

was a new idea. Lee was with the Research Division, and Barnett was with Toll Transmission Division. Barnett's group in Holandel, N.J., and R. Slade's group in Merrimack Valley, Mass., studied Lee's memo and continued his work, but they did not inform Lee. The Bell Lab's rain rate model was published in 1975[13] with no acknowledgment of Lee's work. Internally, Erwin E. Muller, Barnett's boss, wrote to his director, Nathan Levine, who acknowledged that Lee was the first one to create the model (see Exhibit 5.C). However, Lee's own work, "A Simple Method for Obtaining Statistics on Signal Attenuation Due to Rainfall in Major US Cities,"[14] a Bell Labs internal memorandum had been blocked from publication by the Toll Transmission Division until Frank Blecher, Director of Bell Labs, insisted that Lee's work be recognized. Lee's paper was finally published in the *IEEE Trans. on Antenna and Propagation* in May 1979.[15] It was 5 years later and no one paid any attention to this paper anymore.

5.14 THE FAILURE OF THE PICTURE PHONE MARKET

In 1966, Bell Labs invented picture phones that used the same twist-wire telephone lines as used by regular telephones.

Bell Laboratories

subject:

date: July 19, 1977

from: E. E. Muller

Messrs. N. Levine:
        D. G. Thomas:

Several years back when W. C. Y. Lee was originally involved
in rain statistics, some of his work provided impetus for
more sophisticated approaches taken by W. S. Chen, S. H. Lin
and others. Several papers have since been published in this
area with a wrap-up due in the November 1977 BSTJ. Publication
of Lee's paper at this time is largely a means to acknowledge
his contribution. It does not add significantly to the
available reservoir of knowledge. The paper offered appears
hastily drawn; it is not readily comprehensible, it contains
a number of unsupported assumptions.

HO-4363-NFD-kl   E. E. Muller

Exhibit 5.C.   A letter from E.E. Muller.

The twist-wire line can only provide a narrowband transmission. To send picture frames, you need a wideband channel to transmit high-speed data. Their idea enabled a narrowband transmission line to deliver a wideband channel picture of a calling party. At the sending end the picture is first scanned and sent to the receiving end using a slow data rate. After the entire picture frame has been received at the receiving end, only the changing (moving) part of a picture frame is sent. This is a small data stream that the telephone line can handle. An entire picture frame in this case is broken down into many wavelets in a wavelet representation.

The changing part of a picture frame can be identified by certain wavelets that are the only ones to be sent. This means that all the steadily unchanged wavelets are redundant ones and do not need to be sent.

A market survey showed that 70 percent of the interviewers liked the picture phone. When it came out on the market, it cost around $5000 per unit. The Sears Building in Chicago purchased 500 units. But the market did not take off for a fundamental reason. People did not want to call their boss with picture phones. And women did not want to make or receive calls if they weren't looking their best. This type of human behavior was overlooked during the market survey. In addition, for the picture phone to be operable, the other side also has to have one. By contrast, the mobile cellular phone can call not only mobile cellular phones but also any residential phone. The picture phone may have been a very good product, but it failed because it didn't have the flexibility of a cellular phone or the privacy of a nonpicture phone.

5.15   WHY CT-2 FAILED

In 1989, the United Kingdom developed a new cordless phone called CT-2.[16] It was deployed in different markets with different commercial names, such as a zone-phone or telepoint phone system. The CT-2 operates like a portable pay-phone booth. A call can be dialed out but cannot be received (dialed in). CT-2 uses TDD transmission (i.e., the transmitter and receiver share the same channel). CT-2 only has 40 channels and covers a spectrum of 4 MHz. The channel bandwidth is 100 kHz.

In 1986, the Shaye Company, based in the United Kingdom, developed CT-2, and the size of its handset was very impressive. It was the first digital phone. In 1986, even GSM system developers were concerned with CT-2's possible success and were questioning the timing of GSM system's delivery. Unfortunately, Ferranti Co. suggested that the standard proposed by Shaye should be rewritten, which delayed the commercial product. Also, the U.K. government issued four licenses to four CT-2 operators, each one having only 10 channels. Installing 10 channels at each base station could serve about 5 erlangs of traffic. Assuming that the average calling time was 3 minutes, 10 channels could only serve 100 calls per hour. Thus, system capacity with 10 channels was very poor. Besides, the CT-2 phones could only work within the vicinity of CT-2 base stations. Furthermore, four operators could not operate the TDD transmission without a master clock to synchronize all the systems because the filter cutoff was not sharp enough to isolate the interference from the transmitting stage of a system to the receiving stage of the other neighboring system. Poor capacity and mutual interference among the operators made the CT-2 phones lose their markets.

5.16   THE PROS AND CONS OF GSM

The GSM system was developed by a group effort in the European community. All the manufacturing companies agreed to share their IPR. Because of this policy, no one had a reason to keep ideas or inventions private. The GSM development moved very fast. Everyone was participating and making contributions. The morale of GSM's group was very high. Of course, in a group effort the elected leaders might not be the most effective leaders. Sometimes the debating was due to a group member's own self-interests. Then a "compromise" was used to solve the dispute; the result was not always the best technology.

The AMPS analog system was a single standard system that was developed in the United States in 1978 and first deployed successfully in Chicago in 1983. In the meantime the European community felt that it also needed a unified mobile radio system, and GSM group was formed in 1982. At that time, the GSM group was willing to try an advanced and challenging system that had never been done before. The system capacity issue was not brought up at the time because nobody knew that the cellular system would need to be a high-capacity system in 10 years.

GSM reached an early agreement before 1998 that the new mobile radio system would

1. Be a digital TDMA system
2. Have a 300-kHz channel
3. Have 10 time slots

In 1988, a trial of this system was conducted in Paris. The result was very undesirable because the transmission rate for a 300-kHz system was too high and the time delay spread was relatively too long (see Sec. 2.3), so the natural medium wiped out the transmission data. Then it was agreed to lower the bandwidth from 300 to 200 kHz and to reduce the time slots from 10 to 8. But no related theory or experiment appeared in the literature or in any professional journal to prove these values were the optimum solutions. Fortunately these changes to the GSM worked. There were many revisions in the specification, including on the network side. The GSM's success road model was to revise the specification and gradually reach success, as compared with the AMPS success road model, whose first specification was also the final one before commercialization.

In developing GSM, contributions came from all the vendor parties on the team. There was no heroism. In the United States' culture, people are encouraged to invent. The IPR becomes a key drive for inventors, who are treated like heroes in the industry because they can be rich with IPRs. The downside of this IPR is that it could encourage companies to keep their IPR and not openly share with others. Then other companies would have to pay to use IPR technology, forcing other companies to resist using it. As a result, good technology may not be adopted.

5.17   CDI AND CDPD'S TIMING PROBLEM

The Cellular Data Inc. (CDI)[17] system was proposed by Luisgman. It used a 5-kHz band between two 30-kHz channel bands (2.5 kHz in one band and 2.5 kHz in the other band) to transmit low-speed data. It is a spectral-sharing concept. Although the voice quality might be degraded a little, the cellular operator was willing to use it. The CDI system was unsuccessfully developed because of poor strategy. CDI would like to have either GTE or Pactel fund development before it would go full speed in developing its system. Of course, it thought if the operators funded the project, they would commit to use the system when it was commercialized. Actually, CDI should have borrowed the money from banks or venture capitalists and made a successful trial early enough before Cellular Digital Packet Data's (CDPD's) activity started. If it had done so, CDI systems could have been an early cellular data transmission system.

The CDPD[18] system was proposed by PCSI for data service. CDPD found strong support from several major cellular operators led by McCaw Communication in 1992. CDPD also applied the spectral-sharing concept. CDPD uses the existing cellular channels for data transmission while the cellular channels are in idle. When the used channel is suddenly taken by the cellular service, CDPD hops to another idle channel. CDPD took quite a bit of effort and time to develop the frequency-hopping mode. A sniffer device at each base station was used to detect whether each cellular channel was idle or changing to active. However, this device was not effective because the cellular base station had installed a similar device, called foreign interference detector. When CDPD is using a cellular channel, the cellular system will treat that channel as an interfered channel and never use it.

Actually, in rural areas, the traffic is light. There are plenty of idle, unused cellular channels. Therefore, CDPD can have a dedicated channel for data. In urban areas, the traffic is very heavy. There is no idle channel available. If we feel that the data channel is needed, let's have a dedicated channel for data. In any case, CDPD only needs a few dedicated channels, not frequency-hopping channels. An analysis by Lee[19] of using frequency-hopping channels in CDPD concluded that it had a negative impact. Many rivals of CDPD referred to Lee's paper.[20] In a $K = 7$ frequency reuse system, CDPD needs seven dedicated channels, one for each cell. PCSI was interested in developing a frequency-hopping feature because of its involvement in handsets. As a result, developing time was long, the cost of making dual-mode CDPD handset was high, and the frequency-hopping feature could not properly perform in a heavy traffic area. It was clear that the timing of having a CDPD market was lost once the cellular digital systems started to implement the data feature.

5.18   NARROWBAND AMPS[21]

The AMPS system uses 30-kHz bandwidth channels. The capacity of the AMPS system is the same as that of the 10-kHz bandwidth of SSB (see Sec. 3.4). Motorola tried to develop a narrowband FM channel with 10-kHz bandwidth. Then each NAMP channel required a higher C/I (about $C/I = 27$ dB) than the AMPS channel ($C/I = 18$ dB). It had the following relationship,[22] where $(C/I)_1$ is for AMPS and $(C/I)_2$ is for NAMP:

$$\frac{(C/I)_2}{(C/I)_1} = \frac{(BW)_1^2}{(BW)_2^2} = \frac{(30 \text{ kpbs})^2}{(10 \text{ kbps})^2} = 9{\sim}9 \text{ dB}$$

Then, if we just use the same AMPS cells ($K = 7$) and replace the AMPS channels with NAMPS channels, the quality of NAMP channels is poor. Some engineers said that when a portion of 800-MHz spectrum needed to be used for digital system, the NAMPS could gain back the same number of traffic channels as before releasing the AMPS channels. But if we know the reuse frequency concept, we do not need to change to the NAMP system but can simply reduce the K factor from 7 to 3. We regain the same traffic channels.

However, both approaches would lower the voice quality. Changing the K factor does not need to change the system, which would result in a large savings. That is why the GSM system tries to keep the bandwidth unchanged while finding a way to reduce the K factor while maintaining the voice quality. Therefore, the policy is not to change the system, no matter what. The operators should not be pushed by vendors to change the system unless there is an obvious gain.

5.19   MIRS/iDEN SYSTEM

In 1989, Motorola developed a cellularlike system for Nextel,[23] who bought most of the special mobile radio (SMR) spectrum in the 800-MHz paired spectrum (805–821 and 851–867 MHz). It is in a discreted fashion (i.e., not a continuous spectrum). The TDMA system with 25-kHz channel bandwidth and six time slots was called the mobile integrated radio system (MIRS). The equivalent time-slot channel is 4.1 kHz. MIRS used 16-level quadrature amplitude modulation (16QAM), which had four states. Thus 4 times 4.1 kbps equals 16.4 kbps, which might fit in the voice codes.

Using 16QAM modulation can also lower the transmission rate so that the time delay spread doesn't become a problem and the equalizer is not needed. However, there is no free lunch; 16QAM has partial amplitude modulation (AM) and a partial phase modulation (PM). Any mobile radio systems using AM will cause distortion in the voice signal due to the signal fading (see Sec. 2.4).

MIRS was not performing well due to the very narrow band and the effective distortion of the AM part in the 16QAM signal. When Nextel changed MIRS and gave it a new name, Integrated Digital Enhanced Network (iDEN), it only used three time slots instead of six in a 25-kHz band, and the quality got much better. Nextel also used Nortel's switch, which provided great flexibility in implementing features such as offering dispatching and group calls. Although the voice quality is not as good as a cellular system, Nextel's dispatching and group calls feature could gain some customer's preference.

5.20 METRICOM SYSTEM[24]

The Metricom system is a spread-spectrum system using an unlicensed band such as the Industrial, Science, Medical (ISM) band at 2.4 GHz. The data speed is 28.8 kbps, and it uses intelligent nodes called poletops in the radio field to direct calls in both directions. Metricom is a packet data system, its backbone market is the Internet, and it has acquired as many as 30,000 customers. In 1999, the San Francisco area was Metricom's major market. Seattle and New York encompass some of its smaller markets. A Ricochet terminal unit consists of a modem and antenna. Because Ricochet units are used in Metricom's system, the

Metricom's system is also called a Ricochet wireless network. From the poletop to the wired access point (WAP) there are a radio and a router to link the downlink channel, whose bandwidth is 160 kHz, and the uplink channel, whose bandwidth also is 160 kHz. In the future, the system will have a modulation scheme to coop with the data speed higher than 28.8 kbps.

Today, a notebook or laptop computer can be connected to a Ricochet unit and moved to any place at work. A data speed of 28 kbps was still a relatively high speed in the 1999 market. Metricom deployed 5000 poletops in San Francisco, and the cost of each poletop is around $2000. The system is very impressive. Only one question the industry cannot answer is why Metricom does not want to continuously promote its market penetration in San Francisco at this stage. Metricom is instead focusing on the Ricochet^2 Network System at 128 kbps.

5.21   IRIDIUM AND GLOBALSTAR

Iridium is a LEO system (see Sec. 3.9). This high-tech system was designed to cover the globe with 77 satellites. Iridium is named after its chemical element, which has 77 atoms. The number of satellites was later reduced to 66, but the name Iridium was still kept. The signal of a mobile unit sent to the satellite will be switched in space from satellite to satellite and returned to the earth when it reaches its destination. However, because of the high cost of system operation and the lower control capability of traffic flow from any ground stations in the foreign countries, the Iridium system declared bankruptcy in March 2000.

Globalstar is also a LEO system. It covers the globe with 48 satellites. Globalstar is a low-technology system. The signal of a mobile unit is sent to a repeater satellite. The satellite doesn't have a switch to pass the signal from one satellite to another.

When the satellite-mobile service was offered to the international community, most non-European countries were very interested in Globalstar because they can control the satellite traffic and the national sovereignty more easily with Globalstar than with Iridium. This is an example that shows that the advanced technology may not always be the winner. The customer's interests and concerns are more important than the technology's excellence.

Globalstar's system is a relatively low-risk and low-cost system. The cost of each Globalstar satellite is low, but more Globalstar ground gateways may be needed than with Iridium for connectivity.

5.22 LOW-TIER COMMUNICATION SYSTEMS

Low-tier communication systems were suggested in the 1990s when the cost of cellular equipment was still very high. The low-tier communication systems would try to offer low quality with low price service, such as with the personal handy phone service (PHS).[25] PHS was deployed in Japan. There were three operators, NTT, Astel, and DDI. NTT used its phone booths as cordless stations (same as base station) and connected calls through the wire-line backbone. Astel used the electrical power lines for its backbone, which reduced the cost of its system. DDI had to establish its own backbone with microwave links. The systems were started in 1995. In 1998 there were seven million subscribers. However, none of the three companies had earned a profit by that time because the advanced technology reduced the cost of cellular equipment drastically. The cellular service charge also decreased, and the added features, coverage, and mobility made cellular service very attractive. This situation forced the service charge for PHS to be noticeably lower to gain a niche in personal communication services (PCS). Nevertheless, the low end of PCS is the paging service. Between the high end of cellular and the low end of paging, the price margin is not big enough to fit another system such as PHS.

5.23   TIMING ISSUE—SERVICE CREATION STRATEGY

The following several examples illustrate the critical timing in creating new services:

1. Wireless communications can be classified into two kinds of services: fixed and mobile using wireless transmissions. In a building in an urban area, after the wires or cables were laid, adding new wire or cables is very difficult and costly. Using wireless transmission for the fixed services was a better approach; however, in fixed wireless services, WLL service was started after cellular phone service was deployed. The approval of AT&T cellular service took more than 10 years due to AT&T's rival petition to the FCC. As a result, AT&T had to give up and let the seven RBOCs start the service in 1984. If the AMPS system developed by AT&T had been deployed as a WLL system in the 1970s, there would not have been a struggle to deploy the same system as that used in the cellular phone system in the 1980s. If this strategy had been followed by AT&T, the situation of the entire wireless communication industry today might be quite different. The FCC, during the 1970s, might have found that granting a chunk of spectrum for WLL service might be very easy, just as a great chunk of spectrum (54 to 806 MHz) was allocated to the television industry. Later, the same WLL spectrum could have been shared with the cellular telephone systems. For this step, the cellular telephone service could have grown quickly and quietly before the public realized it. This illustrates how critical timing is when creating a new service.

2. In wireless mobile communication, a paging service was not popular in the 1960s and 1970s in rural areas because it needed to use telephones to call paging companies to respond to pages. Without the appropriate telephone systems in rural areas or in developing countries, paging services couldn't be widely accepted.

3. Mobile satellite systems were developed for global wireless communication service and for enhancing cellular or PCS phone services. These systems were developed at the right time in 1996 and would add value to wireless communication as long as the service price was right.

4. The industry trend today is to provide wideband wireless communication services. However, until fiber cables have been deployed nationwide, the development of this service has been delayed because the wireless wideband communication system has to operate at a high frequency, such as millimeter waves or infrared links. Nevertheless, their high propagation loss through the wireless medium is a drawback although the millimeter wave and the infrared technologies can handle the entire wireless wideband communication system. Therefore it has to be a hybrid system, with millimeter wave and infrared links used only for a short range and then using fiber cable for the rest.

5.24   HOW TO SELECT GOOD VENDOR'S EQUIPMENT

Inexperienced operators and engineers may not know how to select a vendor's equipment wisely, especially when it is a newly developed system based on a new standard. Each vendor will quote a price based on how many BTSs, BSCs, and MSCs are ordered. Nevertheless we have to be aware of the following:

1. *The radio quality.*   Each vendor's equipment performs differently. We may need more of one vendor's equipment to have a trial in a given area.

2. *The switch capacity.*   Some vendor's switches have more capacity than others. Also the call processor's capability is very important.

3. *The features.*   Some vendor's equipment has more features than others. If there are fewer features but the cost of equipment is low, we have to ask how much it costs to add those additional features. Only then can we make a good decision about a system.

4. *The redundant and alarm system.*   All cellular system equipment has a redundancy and an alarm system. When one piece of equipment fails, either the redundancy takes care of that piece of equipment or an alarm warns of the situation. It is important to have a good redundant and alarm system; here the higher cost should be justified. Low-cost systems without redundancy should not even be considered.

5. *System maintenance software.*   Good system maintenance capabilities can replace many field engineers. The statistical data obtained at every base station every day can help the system operators tune up the system effectively. A system with poor maintenance software should not be considered.

6. *Penalty clause.*   The penalty clause will be stated clearly in the purchasing contract. If the delivery date or the system performance cannot meet the agreement, the penalty should be applied. But some vendors do not pay a penalty in cash, because they would rather offer credit on the next purchase. It is very unwise for the operators to take this vendor's clause. If the penalty comes from poor equipment performance and credit is given on the next purchase or to the bid of purchasing equipment for another new market, the system operator really suffers for accepting the equipment or the bid based on the low cost. System operators should not be tempted to use more of the undesired equipment to achieve a cost advantage.

5.25   A LESSON FROM PACTEL MICROCELLS

In 1991, Pactel's patented microcell system[26-29] was developed and had a successful trial. Because Pactel had to follow the MFJ restriction and could not manufacture any product, a search for the outsource was conducted. The microcell system needed to be manufactured as quickly as possible due to the need in the market, and Pactel was afraid that a big vendor might be willing to pay a smaller license fee and take a longer time to deliver. Pactel found two good but small vendors to develop Pactel microcells. One was dB Product in Dallas, Texas, and the other was 3dBm in Santa Barbara, Calif. Pactel asked dB Product to produce the zone box (see Fig. 5.12) with optical fiber connectors that could connect to the zone selector at the master zone. Unfortunately, dB Product was sold to Allen Telecom in Cleveland, Ohio. This acquisition delayed the delivery schedule by almost 2 years. If Pactel had thought that this situation could happen, a clause in the agreement could have taken care of it. 3dBm was a small but highly skilled company. Pactel asked it to manufacture a microcell zone box with a 23-GHz and a 18-GHz microwave link for backbone transmission instead of optical fibers. 3dBm's unit was very good except that the company could not stop the rainwater that leaked into the chassis it designed to be mounted on the utility pole. This company did not have the experience to make a weatherproof box. After 3 months, the equipment in the box was damaged by the moisture and performance deteriorated. It took 3dBm almost 18 months to correct it.

Every new innovative system with good technology needs the right timing. Timing is a critical element for success. When the timing is lost, another new innovative system comes out to replace the old one and takes over the market.

5.26 AT&T'S 3B20 SWITCHES

In 1985, when cellular service started, its cost was high. AT&T had already sold its Autoplex 100 switch, which is an analog switch, to all 90 MSA markets for all RBOCs. The non-wire-line (company) operators did not want to use AT&T's equipment for competitive reasons. In this situation, AT&T felt that all the major cellular markets that used its switches would grow very slowly due to the high cost of equipment and the high cost of service at that time. Therefore AT&T targeted small RSA markets for its switches. AT&T developed a small digital circuit switch called Autoplex-10. In 1987, the major cellular operator, Pactel, had already determined that the capacity of Autoplex-100 switches was not enough and searched for a digital switch to carry high traffic. At that time, AT&T could have concentrated on developing the No. 5 ESS large-scale switch for cellular but decided to use the distributed concept to increase switching capacity, as shown in Fig. 5.13. The 3B20 processor was used for the executive cellular processor (ECP) which is a control unit. The distributed system was based on a module configuration. The large digital switch is called Autoplex-1000, which had a distributed architecture and consisted of a number of Autoplex-10s as their modules. AT&T's strategy was that a switch can be customer made depending on the operator's requirement. The token ring configuration could not handle the heavy traffic. Therefore Pactel decided not to use AT&T's Autoplex-1000.

But AT&T did not know that wireless communication would need more capacity for higher traffic and wider band-

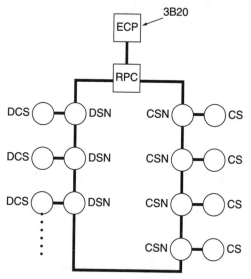

ECP: Executive cellular processor
DSN: Digital switch node
CSN: Cell site node
RPC: Ring peripheral control
DCS: Digital cellular switch (i.e., Autoplex 10)

Figure 5.13.   Cellular switching equipment: Autoplex-1000, a decentralized system

width in the next 15 years. This one decision of using a distributed switching system instead of No. 5 ESS hurt the AT&T cellular system badly. If the large-scale switch had been developed, the rich features and services would definitely have won a large market share. Later, Class 5 ESS was developed, but the loss of the market share could not be recovered.

5.27 SEVERAL IMPORTANT TOOLS FOR NEW SYSTEMS

In designing a system we need many important tools. Without the tools, many key parameters cannot be figured out.

5.27.1 REMOVING THE RANDOM IMPULSES FROM THE DATA

Signaling data collected in the mobile radio field has revealed extremely high values of impulsive noise contamination. A software program is needed to remove high-level impulses from the data. It is not easy to estimate the average of digitized noise data in the presence of high-level impulses. The impulses do not carry any energy. But if caught by the digitizing, the high level of the impulse sample would raise the average noise level. For example, if three impulses are 20 dB higher than the rest of total 256 noise samples, the average level would be raised by 3.3 dB:

$$\frac{253 + 3 \times 100}{256} = 2.16 = 3.3 \text{ dB}$$

Therefore, it is necessary to remove the impulses in the noise data samples before taking the average. Since we cannot easily identify from the digitized data which is impulsive noise, a new technique using the nature of noise statistics to remove the impulses was developed by Lee at Bell Labs.[30] Every piece of noise data had to go through the software program before to find the average noise level.

5.27.2   MULTICHANNEL SIMULATOR

In designing the microcell system,[26–29] we needed a multichannel simulator to simulate a given number of interfering channels, each of which should have 30-kHz channel bandwidth, as AMPS channels do. All the channels, including the undesired channels, would be sent through the microcell system to find the performance at the receiving end. This multichannel simulator was not available on the market. HP did not make it. Usually, only a two-tone generator could be found. Creating multichannel frequencies from a single search was not easy. Lee found a solution and filed a patent.[31] Then, the multichannel simulator was built. It helped to improve the microcell systems in the 1990s.

In May 1995, intermodulation (IM) interference was found in the CDMA mobile receivers when the mobile units were close to the AMPS base stations. The strong base-station signal overloaded the low-noise amplifier (LNA) of the CDMA mobile receiver, and IM components were created and fell into the band of CDMA. As a result, calls were dropped. Because the CDMA band is wider, the chance of having IM components fall into this band becomes greater. To find the solution, C. E. Wheatley and J. Maloney from Qualcomm wanted to seek a multichannel generator to simulate the problem but could not find one on the market. They borrowed the multichannel simulator from Lee's lab and generated 28 AMPS interfering channels (worst-case scenario) in the AMPS band. The CDMA receiver found that 13 IM tones fell in the CDMA band which was next to the AMPS band. These 13 IM tones merely increased the jammer power by 7 dB.[32, 33] One way to reduce the 7-dB interference is to switch LNA out when the received signal at the mobile unit is strong.

5.27.3   APPLY ARTIFICIAL INTELLIGENCE

In 1985, Lee filed a patent[34] for an artificial intelligence (AI) application while he was with ITT Defense Communications Division. At that time AI was a new field. AI is not based on the algorithm approach but follows a set of heuristic rules

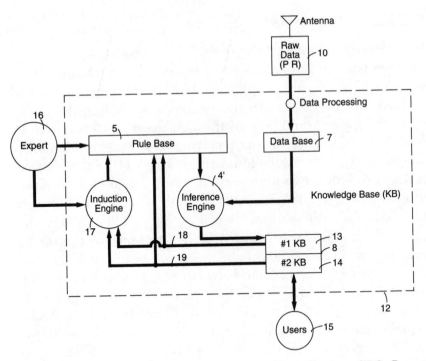

Figure 5.14. Connectivity control by artificial intelligence (U.S. Patent 4,999,833 by Lee).

residing in the rule base, as shown in Fig. 5.14. The rules form a rule basis and the input data from a database. Both the rules and the input data enters the interference engine, which provides a solution or knowledge. It is kept in the knowledge base.

Then there is an induction engine that receives the input from the expert knowledge base to enhance the rule base. This invention was used for the network connectivity control by AI when the master station is destroyed on the battlefield. Unfortunately, in 1985, the patent examiner did not have enough AI knowledge to grant the patent. Therefore, Lee had to supply a lot of AI reading material (see Exhibit 5.D). Lee's AI patent was filed on May 6, 1985, but was not granted until March 12, 1991.

United States Patent [19]

Lee

[11] Patent Number: **4,999,833**

[45] Date of Patent: **Mar. 12, 1991**

[54] **NETWORK CONNECTIVITY CONTROL BY ARTIFICIAL INTELLIGENCE**

[75] Inventor: **William C. Lee**, Corona Del Mar, Calif.

[73] Assignee: **ITT Corporation**, New York, N.Y.

[21] Appl. No.: **287,742**

[22] Filed: **Dec. 20, 1988**

Related U.S. Application Data

[63] Continuation-in-part of Ser. No. 125,738, Nov. 30, 1987, abandoned, which is a continuation of Ser. No. 731,189, May 6, 1985, abandoned.

[51] Int. Cl.⁵ H04J 3/24
[52] U.S. Cl. 370/94.1; 370/58.3; 370/60; 370/94.3; 340/825.02; 340/825.06; 364/513
[58] Field of Search 370/60, 94.1, 58.1, 370/58.2, 58.3, 60.1, 94.2, 94.3; 455/62; 364/513; 340/825.02, 825.06

[56]   **References Cited**

U.S. PATENT DOCUMENTS

4,320,500	3/1982	Barberis et al.	370/94.1
4,601,586	7/1986	Bahr et al.	370/94.1
4,670,848	6/1987	Schramm	364/513
4,779,208	10/1988	Tsuruta et al.	364/513

OTHER PUBLICATIONS

Hayes-Roth et al., "Building Expert System", 1983, pp. 129-131, 287-326.

Primary Examiner—Benedict V. Safourek
Assistant Examiner—Alpus H. Hsu
Attorney, Agent, or Firm—Arthur L. Plevy

[57]   **ABSTRACT**

A communications system utilizes artificial intelligence to select connectivity paths among various locations in a communications network. An embodiment shown is that of a packet radio network, wherein an artificial intelligence module, located at one or more of the radio sites in the network, applies a set of heuristic rules to a knowledge base obtained from network experience to select connectivity paths through the network. The artificial intelligence module comprises an inference engine, a memory for storing network data obtained from a radio receiver and transmitting it to the inference engine, a memory connected to the inference engine which stores a set of heuristic rules for the artificial intelligence system, and a knowledge base memory which stores network information upon which the inference engine draws. The knowledge base memory is also capable of feeding back network information to the rule base memory, which can thus update its rules. Also shown is an embodiment of a multimedia communications network.

17 Claims, 10 Drawing Sheets

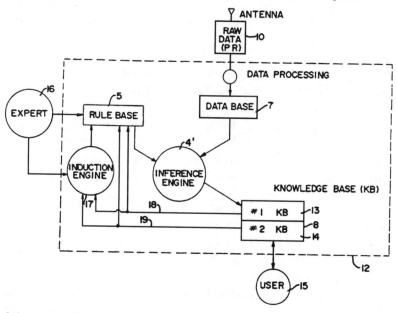

Exhibit 5.D.   "Network Connectivity Control by Artificial Intelligence," patent granted to W. Lee..

5.28   REFERENCES

1. FCC rule making "40 MHz Spectrum Shared with Wireline and Wireless Companies," January 1981.

2. U.S. Court for the District of Columbia "United States of America vs. Western Electric Company, Incorporated, and American Telephone and Telegraph Company—MFJ," Civil Action No. 82-0192, August 24, 1982.

3. W. C. Y. Lee, "Sharing Spectrum and Harmful Interference," *VTC-2000 Spring,* Tokyo Japan, May 15–18, 2000, Conference Record, pp. 1778–1781.

4. B. G. King, "Experiment Using Light Transmission" A letter to Mr. T. F. Sullivan, Maintenance Supervisor, Empire State Building, on December 18, 1972.

5. W. C. Y. Lee, "Measuring Apparatus for Millimeter Wave Propagation in New York City," Bell Labs Memorandum for Record, April 12, 1974.

6. W. C. Y. Lee, "Studying the Advantage of Using a Diversity System Between Millimeter Wave Link and Optical Wave Link in Metropolitan Areas," Bell Labs Memorandum for Record, March 25, 1974.

7. L. H. Von Ohlsen and C. N. Dunn sent the Impatt Diodes from Bell Labs (Reading, Pennsylvania), Bell Labs Internal Memorandum, June 7, 1972.

8. D. A. Gray, "Impatt Diodes," Bell Labs Internal Memorandum, Dec. 11, 1972.

9. W. C. Y. Lee, "Impatt Diode," Bell Labs Memorandum to C. N. Dunn, January 17, 1974, reported on the operating status of 14 diodes.

10. W. C. Y. Lee, "Measuring Apparatus for Millimeter Wave Propagation in New York City," Bell Labs Memorandum for Record, April 12, 1974.

11. W. C. Y. Lee, *Mobile Cellular Telecommunications, Analog and Digital Systems,* 2d ed., New York: McGraw-Hill, 1995, pp. 646–650.

12. W. C. Y. Lee, "No Cost and Fast Time in Obtaining the Signal Attenuation Due to Fog Alone" Bell Labs Memorandum for Record, May 10, 1974.

13. S. H. Lin, "Rain-Rate Distributions and Extreme Value Statistics," *Bell System Tech. J.* 55:1111–1124, October 1976.

14. W. C. Y. Lee, "A Simple Method of Obtaining Statistics on Signal Attenuation Due to Rainfall in Major U.S. Cities," Bell Labs Memorandum for File, June 9, 1975.

15. W. C. Y. Lee, "An Approximation Method for Obtaining Rain Rate Statistics for Use in Signal Attenuation Estimating," *IEEE Trans. on Antenna and Propagation,* AP-27:407–413, May 1979.

16. "Cordless Telephone 2/Common Air Interface (CT2/CAI)," *Management of International Telecomunications,* MIT 12-850-201, Delran, NJ: DataPro Information Services Group, February 1994.

17. CDI System, "The CDI System Specifications," Cellular Digital Incorp., 1994.

18. "CDPD—Cellular Digital Pocket Data, Cell Pat. Plan II Specification," San Diego: PCSI Co., January 1992.

19. W. C. Y. Lee, "Data Transmission via Analog Cellular Systems," *ICUPC Proceedings*, San Diego, Calif., September 27–October 1, 1994, pp. 521–525.

20. Ellen Kayata Wesel, *Wireless Multimedia Communications*, New York: Addison-wesley, 1998, p. 277.

21. Narrow-AMPS, "A Bridge to the Digital Future," CTIA Technology Forum, Chicago, Illinois, Dec. 6, 1990.

22. W. C. Y. Lee, *Mobile Cellular Telecommunications, Analog and Digital Systems*, 2d ed., New York: McGraw-Hill, 1995, pp. 414–417.

23. Graham Haddock, "Nextel Base Station Interference," a Motorola memo to Pactel Co.

24. Metricom, "Metricom's Strategy Based on Huge Number of Users, " *Mobile Data Report* 5(16), August 16, 1993.

25. Personal Handy Phone (PHS), Personal Handy Phone Standard Research Development Center for Radio System (RCR), CRC STD-28, December 20, 1993.

26. W. C. Y. Lee, "Microcell System for Cellular Telephone System," U.S. Patent 4,932,049, June 5, 1990.

27. W. C. Y. Lee, "Small Cell for Greater Performance," *IEEE Communication Magazine*, November 1991, pp. 19–23.

28. W. C. Y. Lee, "An Innovative Microcell System," *Cellular Business*, December 1991, pp. 42–44.

29. W. C. Y. Lee, "Applying the Intelligent Cell Concept to PCs," *IEEE Trans. on VT*, vol. 43, August 1994, pp. 672–679.

30. W. C. Y. Lee, "A Technique for Estimating Unbiased Average Power in the Presence of High Level Impulses," *ICC '80 Conference Record*, 1980, pp. 24.3.1–24.3.5.

31. W. C. Y. Lee, "Frequency Signal Generator Apparatus and Methods for Simulating Interference in Mobile Communication System," U.S. Patent 5,220,680, June 15, 1993.

32. Charles E. Wheatley and J. Maloney, private communication, October 2, 1995.

33. W.C.Y. Lee, *Mobile Communications Engineering*, 2d ed, New York: McGraw-Hill, 1998, pp. 592–594.

34. W.C.Y. Lee, "Network Connectivity Control by Artificial Intelligence," U.S. Patent 4,999,833, March 12, 1991.

APPLICATION OF CDMA

6.1   WHAT IS CDMA?

CDMA is a multiple access scheme[1] that assigns a particular code sequence (which acts like a traffic channel) to the calling party. In each CDMA system, the carrier channel is a broadband radio frequency (RF) carrier, which carries many different code sequences or many different traffic channels. In the early 1980s, the CDMA system was applied to the satellite communication system for transmitting high-speed data. However, the CDMA system was not used for terrestrial mobile systems in the 1980s.

Two popular systems, FDMA and TDMA, can be compared with the CDMA system. The radio structures of FDMA and

TDMA are similar. Each of the FDMA frequency channels and TDMA time-slot channels is analogous to a party of two talking in an isolated room with no interference. A building can be divided into a number of rooms, as shown in Fig. 6.1a. The size of the room is analogous to the bandwidth of the frequency or the time-slot channel. However, the radio structure of CDMA is different from FDMA and TDMA. The CDMA radio carrier is analogous to a great hall filled with people talking to each other, as shown in Fig. 6.1b. No walls isolate the voices so people have to control their voice so that everybody can talk at the same time. A good example is a large formal dining room where people keep their voices low. When entering such a dining room, it is surprising that so many people are in the room because it is so quiet. In a CDMA system, the power control of each traffic channel is very important because all the traffic channels share the same radio carrier channel.

On the other hand, a CDMA system can run like a simulation of a cocktail party.[2] When a cocktail party is held, voices can be heard at a distance. Guests can barely hear each other, so they raise their voices. Soon all the voices in the room are raised and the entire room fills with loud unintelligible noises. As a result, nobody at the party can enjoy a conversation. Without power control, the CDMA system would have the same result as a cocktail party. The CDMA system cannot function properly without power control.

6.2   WHAT IS SPREAD SPECTRUM?

Spread-spectrum (SS) modulation is suitably applied to the CDMA system because a CDMA system uses a particular code sequence to transmit not only data information but also the identified originator and the address code sequence that applies to the data information. For example, an information bit can be represented by a sequence of coded bits, which are called chips. If a bit is represented by 100 chips, the spectrum of information bits needs to be spread 100 times the spectrum of chips. This is called the spread spec-

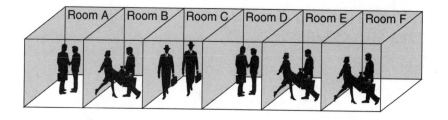

(a) Analogy to FDMA or TDMA - talking in separate rooms

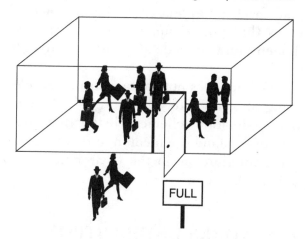

FULL

(b) Analogy to CDMA - talking in a ballroom

Figure 6.1. (*a*) Analogy for FDMA or TDMA: talking in separate rooms. (*b*) Analogy for CDMA: talking in a ballroom.

trum. SS techniques have been used for antijamming of enemy interference since the 1960s. But in cellular CDMA, we are using SS to increase radio capacity. Actually, FM is the earlier SS, invented by Edwin Howard Armstrong. Armstrong realized that by spreading the information signal over a wider band, the ambient noise could be reduced. The FM signal deviation is defined by the FM modulation index, $m = \Delta F/W$, which is the frequency deviation divided by the information bandwidth. For AMPS, $\Delta F = 12$ kHz, $W = 3$ kHz (voice spectrum), and $m = 4$. The FM modulation index may also be treated as the processing gain (PG) of SS modulation. In the cdmaOne system, the baseband data rate is R = 9.6 kbps, and the bandwidth for the chip rate is B = 1.2288 MHz. The PG is

$$PG = \frac{1.2288 \times 10^6 \text{ Hz}}{9.6 \times 10^3 \text{ bps}} = 128$$

Therefore PG = 4 for FM, and 128 for cdmaOne system. The cdmaOne system needs the high PG because the CDMA system uses spread spectrum to reduce the interference and, as a result, increases the system capacity. Thus FM uses SS to reduce the ambient noise and cdmaOne uses SS to reduce the interference.

If the radio medium does not contain strong noise or interference, applying SS is a waste. Therefore in radio communications, no one modulation scheme is better than the others. We have to first understand the medium (i.e., the characteristics of noise and interference in the environment) and then find the suitable modulation.

6.3   WHY SS WORKS UNDER STRONG JAMMING

Under a strong enemy jamming, the jamming level could be 10^5 (50 dB) higher than the receiving signal level, as shown in Fig. 6.2. Then, how can the receiver still receive its own signal? In this section, we demonstrate the value of SS modulation. First, we have to introduce a general equation for digital systems. The received C/I ratio can be expressed by

$$\frac{C}{I} = \frac{E_b \cdot R}{I_o \cdot B} = \frac{E_b/I_o}{B/R} \qquad [6.1]$$

where E_b is the energy per bit, R is the transmission rate (bps), I_o is the noise power per hertz, and B is the channel bandwidth in hertz. The C/I value is obtained at the RF level. It can be greater or smaller than 1. E_b/I_o is measured at the baseband. It is always greater than 1. Usually $E_b/I_o \geq 10$, or 10 dB.

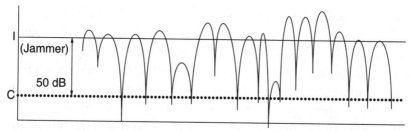

Figure 6.2. The effect of strong jamming on the receiving signal level.

We may assume a condition that the jammer (I) is 10^5 times stronger than the desired signal (C), $I/C = 10^5$. At the baseband, $E_b/I_o = 10$ and the transmission rate is 100 bps. Now under what condition can the SS modulation help in antijamming? Applying Eq. [6.1], and inserting the assumed values in the equation,

$$\frac{C}{I} = \frac{10}{B/100} = \frac{1000}{B}$$

then

$$B = 10^8 \text{ Hz} = 100 \text{ MHz}$$

The answer is that using a SS modulation with a bandwidth of 100 MHz will antijam the enemy interference. The PG of this system is 100 MHz/100 bps = 10^6, or PG = 60 dB.

Figure 6.2 will help explain the physical reason that I and C are the average power values from their instantaneous inputs. Therefore, even though the jamming is very strong, there are many tiny clear time intervals of an instantaneous signal that are not being interfered with. By applying the SS and creating a great deal of redundancy, the weak desired signal can be received.

6.4 CDMA DEVELOPMENT

Qualcomm was a 200-employee company located in San Diego, Calif., that was formed in 1985. Their main business at the beginning was Omnitrak, a system that tracked trucks. In

February 1989, Qualcomm's 10 key people, led by Erwin Jacobs and Andy Viterbi, visited Pactel (now Airtouch). Klein Gilhousen introduced the CDMA system for cellular use to them. They were talking about applying CDMA on the satellite communication to cellular. W. Lee from Pactel mentioned that the unique phenomenon to be considered in cellular was the near-mobile to far-mobile interference, or simply near-far interference. To apply CDMA, the cellular near-far interference problem should be resolved by using the power control scheme in CDMA. The power control schemes for FDMA or TDMA were much easier, and for CDMA it wasn't an easy or obvious task in 1989. CDMA had to control the power of each code sequence within a radio channel. Lee had studied the spread-spectrum system and was granted two patents[3,4] in spread-spectrum communication before 1985. He realized the difficulty of finding a power control scheme for cellular CDMA.

In April 1989, Qualcomm personnel visited Pactel again. They found a solution to power control in CDMA coded channels. After its successful power control presentation, Qualcomm asked for a $200,000 study contract from Pactel. At that time at Pactel, Lee, a corporate technology vice president, suggested to J. R. Hultman, a cellular CEO, and F. C. Farrell, a network vice president, that Qualcomm's paper study would be useless to Pactel. Pactel said it might support Qualcomm by giving it $1 million if Qualcomm was willing to deliver a demonstration of CDMA in 6 months. Qualcomm accepted. The reason for finishing a CDMA demo in 6 months was that the American digital system (TDMA) was already voted as the digital system standard and in the stage of writing its specification. A symposium on digital standards was held and the next generation cellular technologies section was chaired by W. Lee (Exhibit 6.A). Getting CDMA out to the public in a timely manner was very important. Theoretically, CDMA proved that its capacity was at least 10 times that of AMPS. Such a system was what the ARTS of CTIA was looking for (see Sec. 3.6).

Jeff Hultman asked Lee to technically assist Qualcomm's demonstration and to be at Qualcomm at least once a week. Hultman and Farrill were handling other demonstration issues, such as clearing Pactel's frequencies (at least 42 channels) in the Pactel's San Diego Market for a CDMA demonstration. Two Pactel cell sites plus Qualcomm's headquarter site were used for the demonstration.

The CDMA demonstration was a miracle. To purchase any electronic parts in the United States, even a resistor, would take 4 to 6 weeks. How could an innovative system be built from scratch and a demonstration be held in 6 months? The demonstrated CDMA mobile unit and base stations were modified from the existing mobile units. The CDMA switch was modified from the existing PBX, the forward link and reverse link protocols were designed, and the power control was implemented. Almost everything was a first-time experience, and with this highly motivating goal, Qualcomm's engineers worked day and night. It was not surprising to call Qualcomm's lab at midnight and find C. Whitley, B. Weaver, and R. Padovani and their people still there. Good leadership, excellent skills, and hard work made it possible to pursue a dream goal.

In September 1989, Qualcomm asked for another million dollars (in addition to their own investment) to get the demonstration finished in November 1989. Hultman was a little concerned and asked Lee for his opinion during the joint meeting with Qualcomm. Lee answered Hultman's question with a Yes. Qualcomm was very happy, not only because of the $1 million, but also the strong support from a large operator which could be worth more.

On November 3, 1989, the demonstration took place at Qualcomm headquarters in Serrendo Valley. About 60 attendees came, including domestic European, Korean, and Japanese companies. The demonstration was successfully completed, and a flier was distributed during the demo (see Exhibit 6.B). Lee's views of CDMA were printed on the first page of a flier and a video tape was made during the demonstration. It was a shock to the cellular industry, but Qualcomm's demonstration gave the U.S. cellular industry confidence and ensured that U.S. companies had a strong capability to push for technological excellency.

Qualcomm was not a cellular vendor company in 1989, but Lee was known in the cellular industry, and he fully understood that CDMA had a greater capacity than TDMA. In January

Digital
Cellular
Standards

The Next Generation U.S. Cellular Telephone System

A Two-Day Symposium

Sponsored by

The Cellular and Common Carrier Radio Section of the Telecommunications Industry Association and EIA/TIA Standards Committee, TR-45.3

Housing Information

A block of rooms has been reserved at the luxurious J.W. Marriott Hotel (202/393-2000) at a special rate of $89 for a single or double room. The block is under the name of Telecommunications Industry Association Digital Cellular Seminar and has a cutoff date of *August 1, 1988*. Check-in time is after 3 p.m., and check-out time is noon. The J.W. Marriott Hotel is located at 1331 Pennsylvania Avenue, NW, Washington, DC. It is a short cab ride from National Airport and can also be accessed from the Washington Metrorail Metro Center station.

Continental/Eastern Airfare Discount

Continental Airlines and Eastern Airlines are offering special discounts for this event. Coach and First Class fares have no minimum stay or advance booking requirements and no penalty for itinerary changes or cancellations. Call early for the limited special excursion fares at even higher discounted rates.

The convention desk is open Monday–Friday from 8 a.m. to 9 p.m. (Eastern time). Call 800/468-7022 and refer to Easy Access Number EZ8BP68. The convention fare discounts (50% off coach fares, 50% off first class fares, and 5% off applicable excursion fares) are valid for travel from August 13 through August 24, 1988.

Exhibit 6.A. A two-day symposium on the Digital Cellular Standard, sponsored by TIA.

PROGRAM

Thursday, August 18

9 a.m. Cellular Carrier Perspective on New Radio Technologies
Organizer: Dennis M. Rucker, Director of Planning, Ameritech Mobile Communications; and Chairman, CTIA Subcommittee for Advanced Radio Technologies

Topics will include:
• Review & summary of CTIA efforts
• Outline of user's performance specification
• Requirements of a new radio technology: capacities, features, services
• Economic considerations
• Quality of service and compatibility
• What does a new radio technology offer markets smaller than the top 10?

LUNCH

Speaker: Patricia Diaz Dennis
FCC Commissioner

1:30 p.m. International and Regulatory Perspectives
Organizer: Frank L. Roe, Chief, FCC Technical Standards Branch
Representatives from Canada, Europe, and the FCC will address the cellular evolution from their respective points of view.

3 p.m. Overview of Next Generation Cellular Technologies
Organizer: Dr. W. C. Lee, Pactel Cellular
A panel of technical experts will discuss the issues of:
• FDMA and TDMA
• voice coding
• channel coding
• linear and constant envelope modulation
• spectrum efficiency

6:30 p.m. Reception

Friday, August 19

8 am Digital System Proposals
Moderator: Peter Nurse, NovAtel; and Chairman, TIA/EIA TR-45.3 Standards Committee

Following a description of the TR-45.3 standardization process, major equipment suppliers such as AT&T, Ericsson, InM, Motorola, NEC, and Northern Telecom, will present system proposals which will include consideration of important design parameters such as:
• system architecture
• RF and propagation factors
• protocols

Speaker: Richard Notebaert
President, Ameritech Mobile Communications

2 pm Discussion Panel on Digital Proposals for the Next Generation of U.S. Cellular System
Moderator: Jesse E. Russell, Director, AT&T Cellular Transmission Laboratory; and Chairman, TIA Cellular and Common Carrier Radio Section

This concluding session will give attendees an opportunity to question panelists on such issues as:
• intervendor system operation
• network transition
• economic factors
• analog/digital compatibility
• data applications
• roaming
• controlling billing fraud

------------------- Detach Here -------------------

Digital Cellular Technology Symposium
J. W. Marriott Hotel — August 18 & 19, 1988 — Washington, DC

Return to: Suzanne Mullendore, TIA, Suite 400
1722 Eye Street, NW, Washington, DC 20006   202/457/4937

Name: _____ Familiar Name for Badge: _____

Company: _____ Telephone: _____

Address: _____

Cost: $395 from 7/1/88 through 8/12/88; $495 after 8/12/88
Prepayment is required. Written cancellations accepted through August 12.

Payment is enclosed in the form of a _____ Check (payable to TIA)

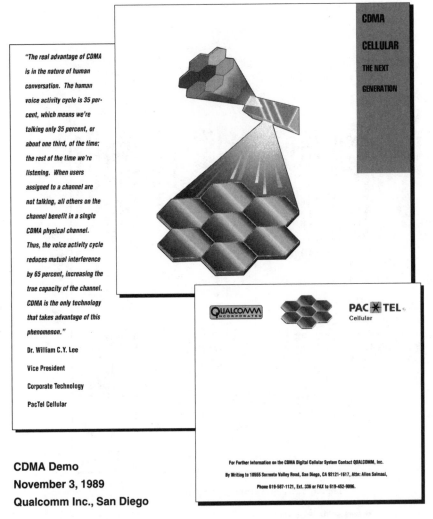

Exhibit 6.B.   CDMA demo flier, Nov. 3, 1989.

1990, Lee took the initiative and gave a 1-day cellular CDMA seminar[5] sponsored by the IEEE West Coast (San Francisco) section and held at the Pacific Bell Auditorium. More than 200 people attended this seminar. In March 1990, Lee went to the East Coast to present the same seminar,[6] sponsored by the IEEE New Jersey section and Dave Goodman's WIN Lab at Rutgers University (Exhibit 6.C). In May 1990, Lee also talked about cellular CDMA at IEEE VTC'90 in New Jersey (Exhibit 6.D).

WIRELESS INFORMATION NETWORK LABORATORY
RUTGERS UNIVERSITY

in participation with the
IEEE Vehicular Technology Society
presents
a tutorial seminar on

Radio Access Technology CDMA/Spread Spectrum
by
Dr. William C. Y. Lee
Pactel Cellular

Date: April 25, 1990
Time: 10:00 AM - 3:30 PM
Fee: $100 ($75 for WINLAB Sponsors)
includes buffet lunch

Place: Sheraton Regal Inn
Kingsbridge Road
Piscataway NJ 08854
(908) 469-5700

Directions to Sheraton Regal Inn: From Route 287S take Exit 5 (Highland Park) and right on to
River Road. From 287N take Exit 5 (Highland Park) and left on to River Road. From River
Road immediately make jug-handle left to Centennial Avenue and second left to Kingsbridge Road.

Registration is limited. To avoid disappointment, please complete the
form below and return before *April 13* to:

Elizabeth Normyle
WINLAB Business Manager
Box 909, Piscataway NJ 08855-0909
Fax: (908) 932-3693
Phone: (908) 932-5954

TUTORIAL: Radio Access Technology CDMA/Spread Spectrum

NAME: _____   AFFILIATION: _____

ADDRESS: _____

PHONE: _____   FAX: _____

Please enclose the $100 Registration Fee. Make check payable to: **WINLAB.**

For WINLAB Sponsors:
Please enclose the $75 Registration Fee and write the words "WINLAB Sponsor" on the check.
Make check payable to: **WINLAB.**

Exhibit 6.C. A tutorial seminar on Radio Access Technology CDMA/spread spectrum at Rutgers University.

DIGITAL CELLULAR TECHNOLOGY PANEL

CHAIRMAN: DR. WILLIAM C. LEE
TIME: 7:30 - 9:30PM
MAY 7, 1990

● OVERVIEW OF NORTH AMERICAN STANDARD SETTING -
DR. PETER NURSE

● SPEECH CODERS -
DR. JAMES MIKULSKI

● TDMA SCHEME -
DR. JAN UDDENFELDT

● CHANNEL STRUCTURE -
MR. JOHN MARINHO

● CDMA SCHEME -
DR. WILLIAM C. LEE

● MODULATION SCHEMES -
DR. KAMILO FEHER

IEEE - VEHICULAR TECHNOLOGY CONFERENCE
ORLANDO, FLORIDA MAY 6-9, 1990

Exhibit 6.D.   The Digital Cellular Technology Panel, IEEE VTC '90.

On Dec. 2, 1990, Lee delivered a paper and participated in a workshop on CDMA at the IEEE GlobeCom Conference, San Diego, California.[7,8]

Lee's four presentations about the cellular CDMA system made a strong wave in digital cellular community in early 1990. Then a special issue of *IEEE Transactions on Vehicular Technology*, edited by Lee, was published in 1991. There were three CDMA papers, Lee's overview of CDMA,[1] K. Gelhousen et al., "On the Capacity of a Cellular CDMA System,"[9] and R. Pickholtz et al., "Spread Spectrum for Mobile Communication."[10]

In the past, each new system required at least 10 years of development before it was commercially viable. For example, AMPS took almost 20 years, GSM took 10 years, and North America TDMA (IS-136) took 7 years. CDMA, however, took only 5 years, making its development the fastest in history.

6.5   THE PHILOSOPHY OF DEPLOYING CDMA

All the traffic channels in a CDMA radio carrier are dependent on each other. They share the total power. When the mobile unit is close to the cell site, it transmits less power to the cell site and lets the mobile unit at the far side transmit more power to the cell site. Every mobile unit considers the others. If one mobile starts to only care for itself and transmit the signal that are as high as it wants, the total CDMA system becomes inoperable. If every mobile unit in the system can control its power more tightly, the interference level in the system is lower and the system can serve more mobile units. Capacity is also further increased, as Lee described in a number of publications.[11–17]

Therefore the game to play is to reduce the interference level in the system. The same analogy is used as in a formal dining restaurant. If the customers can control their voice volume and make it lower, the tables in the dining room can be arranged much closer together. It is the same philosophy as increasing capacity.

Also in the CDMA system, we may find an advantage in changing from the intelligent interference-based background with a few interferers to the unintelligent noise-based background with a great number of interferers. When we are eating in a restaurant, if there are only one or two tables, our talk is interfered with by the other tables. If there are five or more tables, their conversations become unintelligible noise to us, which has much less effect on our conversation.

In a CDMA system, when only a few mobile units are calling, only a few traffic channels are active. During this situation, intelligent interference is high (i.e., a few strong interferers), but the CDMA system has enough room to separate them in the traffic channels. Therefore, the intelligent interference is reduced. When many mobile units are using many traffic channels, the interference becomes unintelligible noise and need not be of concern. The required separation between traffic channels can be much closer. Therefore the CDMA system can be operable in both low- and high-traffic conditions.

6.6   THE ATTRIBUTES OF CDMA

The CDMA system is different from either FDMA or TDMA. We all are familiar with the operations of FDMA and TDMA, especially AMPS, which is a FDMA system. The following table highlights the differences between FDMA, TDMA, and CDMA:

ATTRIBUTES	CDMA	FDMA OR TDMA
C/I	<1	>1
Capacity	Soft	Hard
Handoff	Soft	Hard
Bandwidth	Wideband	Narrowband
Multiple access	Code sequences	Frequencies or time slots
Power control	Tight	Loose
Frequency assignment	No	Yes
Frequency reuse factor (K)	$K \rightarrow 1$	$K > 1$
Voice activity cycle	35–40%	100%
Data transmission rate	High	Low

Some attributes listed in the table are self-explanatory and some are described as follows:

1. CDMA is a soft-capacity system. A CDMA radio carrier is designed to serve from 1 to 55 traffic channels. The number of CDMA traffic channels in real operation is not fixed; it depends on environmental conditions. This is not the same as FDMA or TDMA, where 10 radio channels or 10 time slots can only serve 10 users.

2. CDMA has a soft-handoff operation. In the handoff region, the call always connects to two or more cell sites

to reduce the dropped-cell rate. Also, when the diversity receivers (a multiple of rake receivers) are implemented, the transmitting power of each site to the mobile unit can be reduced. Therefore the sum of the transmitting power received from two or more cell sites in a handoff region can be the same as the transmitting power received from a single site in a nonhandoff region. The radio capacity of this system, in principle, should not be reduced if there is a proper soft-handoff operation.

3. Frequency assignment of each cell in FDMA and TDMA is always a major task because cell sites are always being added from time to time to meet the capacity demand. The macrocells are usually used at system start-up. When the number of cells increase, the size of cells reduce to minicells and to microcells as they become a mature system in an urban area. The frequency assignment in every cell is always changing and becomes a nonstationary process. It is not only labor intensive, but also it's hard to get an optimized solution or even a fair solution in assignment process. The CDMA system does not need frequency assignment. Because one CDMA radio carrier frequency is used in all the cells, the frequency assignment effort in all the cells is eliminated.

4. The human voice activity cycle can be used to benefit the CDMA system. In a CDMA radio carrier, there are many active traffic channels sharing the same radio carrier. Fifty percent is talking time and 50 percent is listening time. Besides, when someone is talking, there are pauses in between. Therefore each user's talking time has a 60 to 65 percent unvoice period. This unvoice period of each user generates no interference to other users. As a result the entire CDMA radio carrier can increase capacity by 2.5 to 3 times.[17]

6.7   THE DARK AGE OF CDMA

CDMA was standardized as another North American digital standard.[18] At that time, the industry treated TDMA as a second-generation system and CDMA as a third-generation system. A IS-95 specification was written in 1991. That same year, most IPR of CDMA was in Qualcomm's hands. Qualcomm was the sole company developing the CDMA system. Motorola, AT&T (later called Lucent in 1995), and Qualcomm were manufacturing the CDMA infrastructures. Oki and Qualcomm were manufacturing CDMA mobile units and handsets. Motorola was the first company to deploy its CDMA system in Hong Kong, in November 1994, and it deployed it in Los Angeles in January 1995. However, Motorola developed the system based on the first issue of IS-95, which was not a mature specification for commercial use in 1995. Furthermore, Motorola's control BSC (CBSC) became a bottleneck in the network under heavy traffic, making it difficult for the operator to increase the number of subscribers while retaining the voice quality of each traffic channel. As a result, the operators could only migrate the heavy users from the analog system to the CDMA system to increase usage but could not increase the capacity. The actual number of subscribers in 1995 was very small. In 1995, U.S. vendors did not have the drive and incentive to push CDMA to success. One of the reasons might be the IPR issue. Qualcomm was a small company, and although it was very skillful with technology, it did not have manufacturing experience. The system operators would rather deal with the big manufacturing companies for large purchase orders. Then, Qualcomm did not have a chance to compete with the major vendors and deploy their equipment first in a market. Without the operation opportunity, it was hard for Qualcomm to improve the CDMA equipment based on the real operation data.

Oki was very active in finding the solution to stopping IM modulation in the handset (see Section 5.27.2) with Qualcomm at the beginning in 1995. Oki's CDMA precommercial handsets were very impressive. However, Oki decided to cease the manufacture of CDMA handsets. Later, Sonney teamed with Qualcomm to produce CDMA commercial handsets.

CDMA's progress was very slow in 1994 and 1995. The articles related to CDMA in magazines and newspapers criticized its technology. The European GSM Community made an unfair comparison. The *Wall Street Journal*[19] even had a negative article about CDMA. Strictly speaking, a successful new communication system requires three factors:

1. The technology should be sound and outstanding among all the other technologies.
2. The manufacturing equipment should be well designed.
3. The deployment system should be well planned and continuously improved.

If one of three is not doing well, the public will blame it on the technology, saying it is no good. In 1991, Lee introduced the CDMA system in China.[20] In 1994, China was in the stages of making a decision on selecting international digital systems. Gao-Feng Zhu, Vice Minister of MPT, made three guiding principles for the Chinese wireless communication deployment:

1. Continue deploying the TACS analog system.
2. Make limited trial of GSM system.
3. Watch CDMA development closely.

Zhu and his staff were very interested in CDMA and visited Qualcomm in 1994. When Zhu could not get information on the growth rate of subscribers from Los Angeles in 1995 and could not to be sure that CDMA was a mature system, he and his successor in late 1995 could not wait and decided to deploy GSM in China. CDMA then lost China's market. GSM ended up with a great and unexpected market penetration in China. At the end of 1997, 1 million subscribers were added every month in China.

During this time period, CDMA's system could not speed up in development and achieve an expected system improvement, as compared with Qualcomm's first achievement in 1989. At the same time, GSM's global penetration drastically increased. The public viewed the CDMA system negatively. This was definitely the dark ages (1994–1996) for CDMA.

6.8   KOREAN CDMA DEVELOPMENT MODEL

In 1990, Lee was invited by the Korean Institute of Communication, through Dr. Hen Suh Park who was Korean Pactel Director at that time, to give a seminar on advanced mobile communications.[21] After the seminar, the Korean President of the Electronic Telecom Research Institute (ETRI), Dr. Sang H. Kyong (who was with Bell Labs and after ETRI, became the Minister of MCC), asked Lee how to develop advanced mobile communication equipment in Korea. Developing a new communications system requires five skills:

1. Developing switch capability
2. Developing chip capability
3. Developing radio capability
4. Developing software capability
5. System integration capability

Korea at that time already had the basics of these five required skills. Lee, from Pactel, suggested technically not the TDMA system but the CDMA system to Korea and spent several hours presenting the CDMA system. The Korean government agreed with Lee's technical suggestion and approached Qualcomm. In November 1990, Lee, from Pactel, and Alan Salmasi, from Qualcomm, again gave the Korean government a formal seminar on the CDMA system.[22] On the same trip Korean ETRI signed an Agreement to purchase the technology transfer of the CDMA package. Another three-day seminar was conducted early the next year.[23] Also, the government announced that the CDMA system would be their only national standard for digital cellular systems and that the CDMA system's manufacturing product should be localized in Korea.

During their CDMA developing stage, the analog cellular market in Korea was growing very fast. The voice quality was getting poorer due to the overcrowded traffic situation, and the digital cellular licensees had acquired their license in 1994. During the Korean license application, both Pactel and Southwestern Bell, from the United States, claimed to be CDMA technology pioneers. To prove that Pactel was the first involved in CDMA, Pactel had to show a letter from Erwin Jacobs to Pactel's president of international operations, Jan

5455 Lusk Blvd, San Diego, California 92121-2779   □   (619) 587-1121   □   Fax (619) 452-9096

January 31, 1994

Mr. Jan Neels
President & CEO
Pacific Telesis International
2999 Oak Road, MS 1050
Walnut Creek, Ca 94596

Dear Jan:

This brief letter is to express my appreciation for the long-term commitment and leadership which PacTel Corporation has extended in the progression of CDMA from a brilliant idea to a powerful reality.

From the beginning in 1989, PacTel has supported our work on CDMA in a wide variety of significant ways. As the first cellular operator that QUALCOMM worked with, PacTel provided us with the experimental FCC license, the cell sites and switching facilities, and funding to build a demonstration CDMA system. PacTel supported the first-ever November, 1989, field demonstration with a group of its San Diego engineers and technicians and attracted over 250 participants from around the world.

During these early stages, your Vice President and Chief Scientist, Dr. William C. Y. Lee worked with us to assist in translating the results of our research and development into a technically precise and workable Common Air Interface (CAI). PacTel also at that time gave us not only local on-site technical support but also the benefit of its years of experience as a cellular operator to effectively apply CDMA technology to terrestrial cellular networks.

In early 1990, PacTel encouraged us to expand the cellular industry support for CDMA technology and by the third quarter, Ameritech, NYNEX, Motorola and AT&T Network Systems had signed funding and licensing agreements with us.

In 1991 and 1992, PacTel and QUALCOMM continued to jointly Field Test this first-ever CDMA network using our technology under actual operational conditions in PacTel's commercial San Diego cellular system. The results of these trials referred to as CAP 1 and CAP 1.1 assisted in the further refinement of the technology and allowed us to get rapid feedback in the development cycle, ultimately producing a field-proven CAI technical standard.

Exhibit 6.E.   The letter of Dr. Irwin M. Jacobs to Mr. Jan Neels (page 1).

January 31, 1994
Mr. Jan Neels
Page 2

Internationally, we particularly appreciated the support of Dr. Hen Suh Park, Pacific Telesis Korea's Representative Director, who introduced us to Korea's ETRI and major manufacturers in 1991. This enabled Korean industry to take the lead in developing and manufacturing the CDMA infrastructure worldwide.

Concurrently PacTel was working in the regulatory and industry environments to obtain the FCC experimental license required for testing this new technology and later to spearhead the drafting and acceptance by the industry of the CDMA standard. We are particularly appreciative of PacTel's leadership and support efforts within the Cellular Telecommunications Industry Association (CTIA). Thanks to the Wide Band Spread Spectrum initiative lead by PacTel in January, 1992, open forums were conducted on the need for a wideband digital cellular standard and the support of industry players was developed, leading to an affirmative vote by the CTIA in June, 1992, moving CDMA in the standards process. PacTel's diligent efforts in the Telecommunications Industry Association (TIA) contributed to the rapid development and consensus around the IS-95 CDMA standard, published in just under 13 months.

In our judgment, PacTel has contributed more actively to making CDMA a commercial reality than any other carrier in the cellular industry.

At present we are teamed up on evaluating CDMA voice quality through extensive testing on the San Diego system and have developed some results that I believe you will be most interested in. (I have attached a copy of our press release on the testing.) We are greatly encouraged by the results of these tests which indicate that ninety percent (90%) of the existing cellular users participating in the test found CDMA cellular service better than or equal to their current analog service.

We look forward to pioneering new applications of CDMA technology to wireless businesses with PacTel in the future. Please feel free to call on us if we be of further assistance.

Very truly yours,

Dr. Irwin M. Jacobs
Chief Executive Officer
Chairman of the Board

QUALCOMM

Exhibit 6.E.   The letter of Dr. Irwin M. Jacobs to Mr. Jan Neels (page 2).

Neels (Exhibit 6.E). Upon receiving their license, the licensees asked the government if they could buy the CDMA system from the U.S. companies to deploy the CDMA system in Korea earlier, without waiting for the Korean product, whose commercialization date they could not even predict. The Korean government rejected this request. Also Korean Telecom (KT) allocated a band of frequencies for a digital cellular system at that time. KT realized the CDMA development was not going smoothly and received an offer from Ericsson in which the GSM equipment could be deployed in Korea without initial payment. KT asked for, but did not get, government approval.

In Korea, engineers at ETRI led by Dr. Seung-Taik Yang and Dr. Hong-Gu Bahk, as well as the engineers within the communications industry such as those from Samsung, Hyundai, and Lucky Glodstar (LG), were working together very hard. However, the CDMA development task remained extremely difficult.

In January 1996, the local company's infrastructure equipment was deployed in Korea's two operator markets. At the beginning, after turning on the CDMA system, the voice quality was very unacceptable, and the dropped call rate was very high. But by April 1996, 3 months later, the performance of the system improved. The operators decided to go commercial. The operator's field engineers and the vendor's engineers worked together, reviewed the field data at the end of each day, modified the system, changed the system parameters or algorithms at night, and tested the system the next day. The subscribers kept increasing, yet the system performance got better. In November 1996, Korea's system exceeded 1 million CDMA subscribers. The CDMA system was no longer a toy. Korea proved that the CDMA system was a high-capacity cellular system. Korea, which did not have any past experience and background in mobile communication systems, in a very short period of time became a major player in global advanced system manufacturing industry. We should learn from its drive and wisdom.

6.9 QUALCOMM INVENTED CDMA AND KOREA RESCUED CDMA

By the end of 1996, the Korean market had exceeded 1.3 million. Although the country had no wireless communication equipment background, but took Qualcomm's transferred technology, believed in it, and developed the system by itself. Korea found many innovative ways to solve practical problems. In cellular systems, a new system that can be operable today may not be good for tomorrow because of increasing users in the system. Therefore, improving the system is a dynamical process. As mentioned in Sec. 6.7, the third required factor for a system to be a success is that the deployment of a system should be well planned and continuously improved. Korean engineers went through this stage and should be proud of their achievement.

When Lee, from Pactel, and Alan Salmasi, from Qualcomm, went to Korea in 1991, they knew that CDMA was an advanced and excellent technology. But it was also a risky technology for Korea. U.S. companies were already starting to develop the CDMA system. They hoped that in the future, Korea would learn from the U.S. development in CDMA. When the CDMA system ran into its dark age, no U.S. company could find a way to break this unhappy situation. Also, U.S. companies did not work as hard as they should have. The vendor's engineers who worked at the CDMA deployed area quit at five o'clock each afternoon, leaving the system operator's engineers to try and figure out the unsolved problems by themselves. Korea's success in CDMA at the end of 1996 shocked the world. No one would doubt the technology of CDMA anymore. At this time, U.S. companies had gained the confidence to pursue CDMA development. CDMA became a great technology. In the meantime, two companies from two non-CDMA regions, Ericsson and DoCoMo, started to call for developing a third-generation (3G) wireless digital system. The year was 1997 and no system operators were looking for 3G in that year. Operators were only studying the new technologies of increasing the capacity from existing systems. CDMA's success did trigger the GSM or TDMA vendors to urgently seek a better solution for their systems.

Besides, the high-speed data transmission in the wireline systems came up rapidly because of the popular Internet service. The future wireless 3G will have a great reason to support high-speed data transmission in the wireless communication world.

Every new activity afterward for the 3G development was more or less triggered by Korea's success in developing CDMA toward a mature high-capacity system.

6.10   CHOOSING CDMA SYSTEMS FOR 3G

In 1990, after CDMA had been developed, Jan Udenfelt, from Ericsson, met with Qualcomm, but no further actions were taken. Ericsson then decided to go full speed into pursuing the GSM markets. If CDMA's development had been completed 2 years earlier, Ericsson's decision might have turned out differently. Nevertheless, Ericsson, Nokia, and other European vendors penetrated the GSM system in the global markets very rapidly. Until 1997, GSM had a 70 percent global market. But in 1994, Jan Udenfelt invited Lee to visit his department research activities at Ericsson. Ericsson had quietly studied CDMA technology themselves but did not mention it in public. To criticize CDMA technology and confuse the U.S. investment community was the GSM vendor's belief. In fact from 1995 to 1996, the performance of the CDMA system was too poor for anyone to defend it anyway.

In the European communities, ever since the CDMA system was developed in the United States, the ETSI was engaging in its next-generation system study. One organization was the Universal Mobile Telephone System (UMTS) group. As mentioned in Sec. 4.11, there were five study groups: TDMA/sync, TDMA/async, CDMA, OFDM, and ODMA. Each group studied its technology independently. The debates and discussions were among themselves. The Korean CDMA system demonstrated its capacity superiority and helped their study. In January 1999, UMTS chose the CDMA for its 3G system and went with the cdma2000 system proposed by the United States. Now the global standard is CDMA. But as shown in Fig. 4.3, there were 13 modes of CDMA from eight standard bodies in the world. Therefore, the 13 modes had to be reduced to three, as discussed in Sec. 4.11. The CDMA radio technology overview of the three modes is covered in Sec. 4.12.

6.11   GLOBAL 3G SPECTRUM ISSUE

Now the global 3G spectrum becomes an issue. In 1994, the ITU had allocated the following spectrum for ground mobile systems:

1885 to 2025 MHz (140 MHz) for mobile transmit

2110 to 2200 MHz (90 MHz) for base transmit

and allocated the following spectrum for mobile satellite systems:

1980 to 2010 MHz (30 MHz)

2170 to 2200 MHz (30 MHz)

The spectrum allocation is shown in Fig. 6.3. Also, the spectrum was allocated from different regions of the world.

In 1996, the U.S. FCC auctioned PCS bands that are in the spectrum ranges from

1850 to 1920 MHz (70 MHz) for mobile transmit

1920 to 1990 MHz (70 MHz) for base transmit

as shown in Fig. 6.3. Now, the base transmit band of the U.S. PCS has fallen in the mobile transmit band of IMT-2000. This means that even though the two bands (IMT-2000 and PCS) overlap, the PCS band cannot be used for IMT-2000 systems. The IMT-2000 mobile transmitter would interfere with the PCS mobile receiver. For this reason, the 3G standard groups are suggesting new bands. The 3G is concerned with the use of the frequency bands 2500 to 2580 MHz and 2610 to 2690 MHz as possible global extension bands. However, these frequencies in the United States have been assigned to the Metropolitan Multipoint Distribution Services (MMDS), which is mainly for WLL services (see Sec. 7.16). Therefore, the agreement from the OHG suggested that the ITU find a future 3G spectrum band below 3 GHz. A common global 3G spectrum is needed so that it can be used for global roaming.

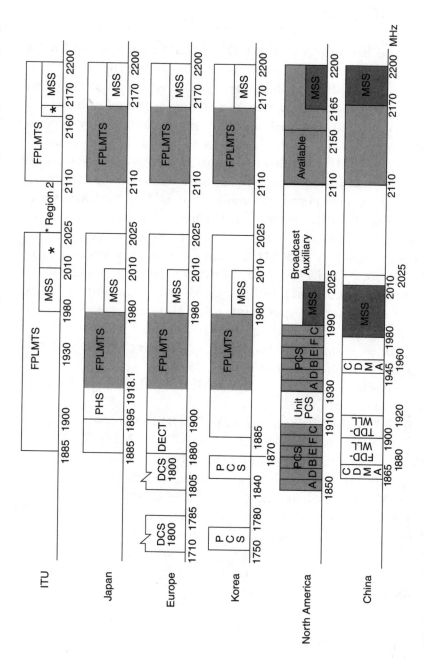

Figure 6-3. Diagram of global spectrum allocation.

6.12   SINGLE CARRIER VERSUS MULTICARRIER

What the bandwidth of the CDMA radio carrier should be is another question to be answered. In the beginning, some proposals suggested 10 or 20 MHz,[24] but most proposals suggested that the technology was ready for the wideband over 5 MHz. Also, the radio medium might not be suitable for a bandwidth over 5 MHz. But, there was no proof of this. Besides, allocating 10-MHz bandwidth channels in a global (IMT-2000) spectrum was harder than allocating 5-MHz bandwidth channels in the same spectrum. Anyway, the 5-MHz bandwidth was decided upon easily without any theoretical justification.

The WCDM, or also called FDD-DS developed in Europe, has only a 5-MHz single carrier system, but the cdma2000, called FDD-MC, has two versions, cdma2000 $1x$ (1.25 MHz) and cdma2000 $3x$ (5 MHz). From point of view of the utilization of spectrum, cdma2000 has a flexible use of spectrum depending on the traffic requirement. Sometimes, cdma2000 $1x$ would be sufficient to handle the traffic; then the spectrum of 1.25 MHz should be used. There is no need to waste the 5-MHz band if the 1.25-MHz spectrum is sufficient.

In 1999, the Chinese version of TDD mode also was standardized from its narrowband version. The 5-MHz TDD carrier can be divided into three narrowband carriers, each one being 1.6 MHz. The chip rate of the narrowband carrier is 1.288 Mcps, which is one-third of the standard chip rate of 3.84 Mcps in a 5-MHz carrier. In this case, two versions of TDD mode can be chosen for different traffic requirements. Thus, TDD is also a multicarrier system. It appears that the multicarrier specification has merit for use in 3G for great flexibility of spectrum utilization.

6.13   REFERENCES

1. W. C. Y Lee, "Overview of CDMA system," IEEE transactions on Vehicular Technology, Vol 40. May 1991, pp. 303–312.

2. W. C. Y. Lee, "Mobile Cellular Telecommunications," 2nd ed. McGraw-Hill Co. N.Y. 1995, p. 575.

3. W. C. Y. Lee, "Covert Communication System," US Patent Office, Patent No. 4,607,375, Aug. 19, 1986.

4. W. C. Y. Lee, "Digital Hopped Frequency, Time Diversity System," US Patent Office, Patent No. 4,616,364, Oct. 7, 1986.

5. W. C. Y. Lee, "The Third Generation of Cellular System–CDMA," sponsored by IEEE West Coast Section/Pacific Bell, Pacific Bell Auditorium, San Ramon, Calif., March 1, 1990.

6. W. C. Y. Lee, "Radio Access Technology CDMA/Spread Spectrum," sponsored by IEEE New Jersey Section and Rutgers University, Piscataway, N.J., April 25, 1990.

7. W. C. Y. Lee,. "Implications of CDMA for Cellular System Operations," *Globecom'90, Workshop 2,* December 2, 1990, Sheraton Hotel, San Diego, Calif.

8. W. C. Y. Lee, "Overview of Cellular CDMA," *1990 IEEE Globecom Conference,* December 2, 1990, San Diego, Calif.

9. K. Gelhousen, I. Jacobs. R. Padovani, A.Viterbi, Le Weaver, C. Wheatley, "On the capacity of a Cellular CDMA System, IEEE Transactions on Vehicular Technology," Vol. 40, May 1991, pp. 303–312.

10. R. Pickholtz, L. Milstein, D. Schilling, "Spread Spectrum for Mobile Communication, IEEE Transactions on Vehicular Technology," Vol 40, May 1991, pp. 313–322.

11. W. C. Y. Lee, "Power Control in CDMA," *IEEE VTS'91* Conference.

12. W. C. Y. Lee, "Cellular CDMA," 1991 *IEEE International Solid-State Circuits Conference,* February 14, 1991, San Francisco.

13. W. C. Y. Lee, "Getting Down to the Nitty-Gritty of CDMA," *Telephone Engineer & Management,* vol. 95, no. 9, May 1, 1991, pp. 72–79.

14. W. C. Y. Lee, "CDMA—An Alternative Approach to Digital Cellular," *International Mobile Communications 1991 Proceedings of the Conference,* London, June 1991.

15. W. C. Y. Lee, "Application of CDMA to Personal Communications Systems," *Fifth Annual Communications Update, Vehicular Technology Society Seminar,* New York, June 28, 1991.

16. W. C. Y. Lee, "CDMA Today," *RNT Magazine,* San Paolo, Brazil, July 1992.

17. W. C. Y. Lee, "A Description of Voice Activity Cycle and the Advantage of Using CDMA," Qualcomm flier/Pactel Demo, November 3, 1989.

18. TIA TR45.5, assigned the CDMA specification as IS95, IS97, IS98, in 1993. TIA/EIA/IS-95, "Mobile Station—Base Station Compatibility Standard for Dual-Mode Wideband Spread Spectrum Cellular

System", Telecommunications Industry Association (TIA), July 1993; TIA/EIA/IS-97, Base Station Minimum Performance Spec., TIA, 1993; TIA/EIA/IS-98, Mobile Station Minimum Performance Spec., TIA, 1993.

19. Wall Street Journal "Jacob's Patter: An Inventor's Promise Has Companies Taking Big Cellular Gamble," by Q. Hardy, September 6, 1996

20. W. C. Y. Lee, "What is CDMA" a 3 day seminar held at Designing Institute, MPT, Zheng Zhou, Henan, China, June 3, 1992.

21. W. C. Y. Lee, "Advanced Mobile Communications," Korean seminar held at Han Yang University, Seoul, Korea, August 6–8, 1990.

22. W. C. Y. Lee and Alan Salmssi, "Digital Cellular," held at Electronics and Telecommunications Research Institute, Daejeon, Korea, Nov. 29, 1990.

23. W. C. Y. Lee, "Mobile Cellular Telecommunications Systems," Korean seminar, Seoul, Korea, April 1–3, 1991.

24. U.S. WIMS "A Broadband CDMA" a proposal submitted to ITU, Geneva, Switzerland as a 3G system candidate in 1998.

WHAT IS OUR FUTURE?

7.1   FINDING A HOME FOR THE GENIUS

Geniuses have unique characteristics. They are quick thinkers and when they are engaged in a conversation about their area of expertise, they tend to be too impatient to explain their ideas clearly and often find themselves having difficulty expressing all of their ideas. As a result, people don't understand what the genius is trying to convey. Instead, it seems as if the genius is just rambling unintelligibly about one idea and then segueing into another topic before the other was fully explained. For the listener, the result is a jumble of disjointed ideas and concepts. The genius is completely misunderstood and is sometimes mistaken for being scatterbrained and eccentric and can be treated as retarded if his or her talent cannot be recognized.

Geniuses' research papers also elude readers because most of their notations and terminology are far too unconventional for most engineers to decipher. Professors at prestigious universities have no trouble recognizing geniuses in their classrooms, and when one is discovered, the genius can perform well in an academic environment. Once geniuses graduate, however, they are faced with the harsh reality that being intelligent doesn't provide them with the ability to open doors in the business world.

Today's corporations cannot provide the appropriate working environment for a genius. A genius needs a comfortable place to conduct research and create inventions. Geniuses are not like other employees; they prefer to work alone. It is not unheard of for these great minds to spend as long as 2 years turning an idea into something concrete. Can any of today's corporations afford to keep these people for such a long period? At the end of those 2 years, a genius might emerge from his or her office with a novel-sized stack of papers, complete with diagrams, figures, equations, and other writings. The challenge is finding someone who can understand these research findings. A capable corporate research head is needed to competently understand the material, otherwise a genius' valuable work is not recognized.

Bell Labs was once a haven for the genius in the old days of Ma Bell. During that time, Bell Labs received 1 percent of AT&T's gross income for their research expenses. In 1960, 1 percent was equivalent to about $1 million tax free per day. Bell Labs in its heyday gave its geniuses an excellent working atmosphere that was both supportive and appreciative. Many great inventions and theories were realized at this time. But once AT&T was divested, few companies wanted to devote the time, energy, or budget solely to research. If industries can't make research and inventiveness a priority, the government should take this responsibility and provide a home for one of the nation's most precious resources—the genius in science and engineering.

7.2 G3G AND ITS HARMONIZATION

The difference between each wireless communication generation is its radio structure. Two or three different generations can be operated at the same time. Using low-speed and high-speed data transmission to distinguish between two generations is very improper. The analog systems are the first generation and use FDMA. GSM, NATDMA, and PDC are the second generation and use TDMA. cdmaOne is the third generation and uses CDMA. Some communities called the cdmaOne system 2.5G, which is not fair; it is completely different from 2G.

G3G (global 3G) was proposed for the first time by DoCoMo and Ericsson as a global standard system. There are large benefits for customers, operators, and manufacturers from a harmonized G3G standard:

1. Provides the best possible "Internetlike" growth trajectory for the mobile industry
2. Maximizes investment in devices and applications
3. Minimizes investment risk in 3G systems
4. Lowest potential cost of devices and services
5. Greatest consumer confidence in the product
6. Best outcome for the ITU development sector
7. Easiest for integration with IMT 2000 satellite component

The agreed-on requirements for G3G are as follows:

1. Global roaming
2. Nonvoice application (high-speed data)
3. One standard and open interface standards

4. A bandwidth of 2×5 MHz of uncoordinated spectrum

5. Backward compatibility

6. Low cost to commercialize and operate

Based on what was learned from cdmaOne, G3G has adopted the CDMA system worldwide (see Sec. 6.10).

Thirteen proposed systems (see Fig. 4.3) were submitted to ITU. It was very difficult to converge from 13 modes to one system, so the harmonization of G3G began. The challenges of harmonization are

1. *Different starting points.* For instance, GSM deployment started in 1992, PDC deployment started in 1993, and cdmaOne deployment started in 1995.

2. *Vendor rivalry.* Competition would remain among themselves.

3. Pride and politics in each region of the world.

4. Lack of consistent regulations in every country.

5. Majority of operators are silent.

6. Lack of focus for customer/industry benefits.

7. 2G competitive environment.

8. Intellectual property.

These challenges are compounded by lack of effective understanding and communication. The current major cellular systems would be as follows:

PDC → CDMA
GSM → GPRS → EDGE → CDMA
IS-136 → EDGE   → CDMA
cdmaOne → 1xRTT   → CDMA

The radio part of G3G was harmonized into three modes that were agreed upon by the G3G community. The harmonization effort is described in Sec. 4.11. Two modes are in the paired frequency spectrum, and one mode is in the unpaired frequency spectrum:

1. *Paired frequency spectrum used for a FDD system*:
   a. *DS.*   A direct sequence in the 5-MHz band, one carrier in the 5-MHz band.
   b. *MC.*   A direct sequence in both a 1.23- and a 5-MHz band frequency spectrum.

2. *Unpaired frequency spectrum used for a TDD system*:   A time division duplex system used in an unpaired spectrum band that benefits the unsymmetrical traffic transmission. The forward link transmission rate can be high, and the reverse link transmission rate can be low. Using TDD has its limitations. With it, the size of cells would be smaller, and the speed of mobile units would be slower. In the future, a breakthrough technology will be found so that a TDD system can be used for a large-area, high-mobility system to compete with FDD systems.

Figure 7.1 shows the global standard vision of three modes.
    The physical layer (Layer 1) is the radio access. As of

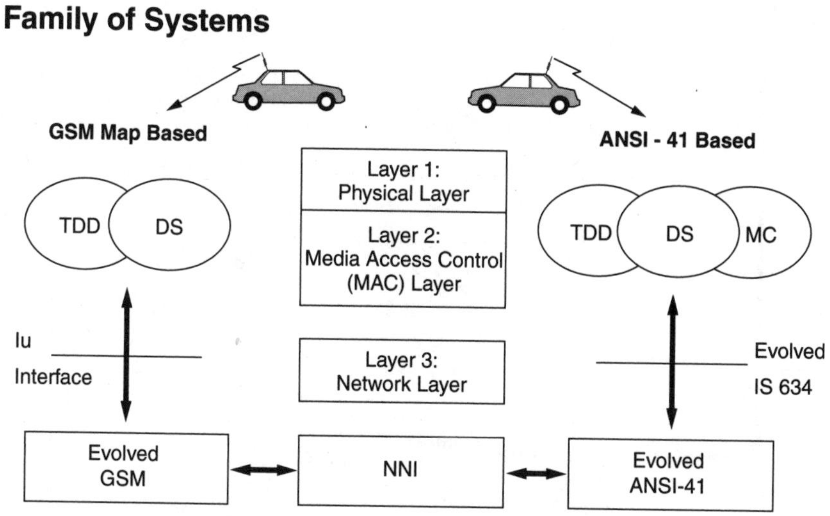

Family of Systems

Figure 7.1.   Global standard vision.

today, the media access control layer (Layer 2) can have either GSM Media Access Protocol (MAP)–based or ANSI-41–based protocol standard. The two control layers can be implemented in each of three modes in G3G and are able to connect through the NNI in the network layer (Layer 3).

The G3G harmonization of the GSM and cdmaOne systems would go through the path shown in Fig. 7.2.

At the end of 1998, during the London G3G conference, the operators felt that G3G would have a great impact on the customers and operators. Since then, Operator Harmonization Group (OHG), an ad hoc organization, was formed and several international operator's conferences have been held (see Sec. 4.11). The harmonization process in each meeting was as follows:

7.2.1   BEIJING MEETING

The outcome of Beijing G3G operator's harmonization group workshop were

1. *Harmonization framework.*   The three modes DS, MC, and TDD were agreed on by the international operators.
2. An IPR open letter was sent to the ITU, the Radio Communication Bureau, and TG8/1.

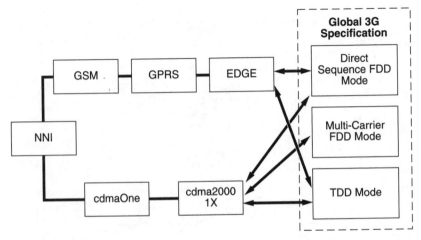

Figure 7.2.   3G harmonization.

3. The technical parameters would be conditional upon resolution of the following issues:

    a. Chip rates between two FDD modes
    b. Pilot structure
    c. Synchronous/asynchronous mode

7.2.2 LONDON MEETING

The outcome of the London G3G operator's harmonization group workshops is shown in Fig. 4.4; the middle column shows the tentative agreement among the operators. For G3G, the multiband spectrum needs to be considered as shown in the following table:

SPECTRUM (MHz)	SPECTRUM BANDWIDTH	MHz	INBAND SERVICE	POSSIBLE MIGRATION
800	50	Subtotal	AMPS,NAMPS, CDMA, TDMA	3G
900	50	100	GSM	3G
1500	48	148	PDC	3G
1700	60	208	Korean PCS	3G
1800	150	358	DSC-1800	3G
1900	120	478	PCS	3G
2100	$2 \times 60 + 35$		Future 3G	3G
Total	633			

Today's existing band has 478 MHz, which is 76 percent of the total available band. Therefore the utilization of spectrum in the existing bands for G3G is essential.

The chip rate is a transmission rate used in CDMA to distinguish it from the bit rate. In cdmaOne, the bit rate of a traffic channel is 9.6 kbps. The chip rate is 1.2288 Mcps. We may say 1 bit is spread into 128 chips. The chip rate issue in G3G is a symbolic issue and has nothing to do with the IPR issue. We can illustrate the negligible effect of different chip rates based on 3.84 Mcps as compared to the effect on other technologies (see Fig. 7.3). Introducing smart antenna technology into G3G would provide a 6-dB gain, or an increase of 400 percent in capacity. Introducing a new vocoder EVRC or AMR would provide an increase of 4 dB, or 250 percent of its capacity. Introducing improved handoff algorithms would provide an increase of 3 dB, or 200 percent of its capacity. The capacity due to the different chip rates becomes relatively insignificant. The 4.096-Mcps chip rate has a 6 percent increase in capacity, and the 3.68 Mcps has a 4 percent decrease in capacity as compared with 3.84-Mcps chip rate. If the operators can agree on the 3.68-Mcps chip rate as the single standard chip rate, the cdmaOne system will benefit because its chip rate is 1.2288 Mcps; 3 times that chip rate is 3.68 Mcps.

Using CDM or TDM as the pilot structure has different benefits. With CDM, the channel estimation (average power) information can be collected within the signal stream. Using TDM requires that channel estimation information be collected from another signal stream. But the overall pilot power may be reduced.

The use of synchronization (sync) versus asynchronization (async) in G3G is another debate issue. The sync system is

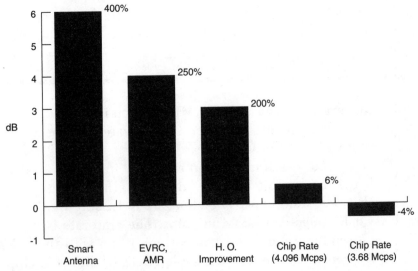

Figure 7.3. Effect on different chip rates based on 3.84 Mcps with other technologies.

actually a subset of the async system. Without a master clock in space to feed to the base stations all the time, async is the only way to handle synchronization from the beginning. An async system starts from one master clock at one location (one of the base stations) then synchronizes on a per-call basis at each base station. The synchronization system would better use the global position satellite (GPS) and eliminate the step from async to sync. In a soft-handoff region, there are often four to six pilot signals coming from the adjacent base stations. Then, making async-to-sync change for each signal takes time. If a strong signal does not receive its sync quickly, it will become a strong interference and cause the call to be dropped. Sync can have less latency and less chance of dropping calls than async in soft-handoff region.

The following was learned from G3G harmonization effort:

1. Clearly separate political and technical issues.
2. Understand the culture of each region:

   a. In the United States, use IPR to create an innovative product.
   b. In Europe use a collective effort to achieve common standards.

3. Harmonization requires a willingness to make some degree of concession.
4. Operators want to select the best technologies, yet have low cost, low risk, and high performance.

7.2.3 TOKYO MEETING

The outcome from the Tokyo workshop was as follows:

1. Chip rate:

   a. WCDMA (DS) uses a chip rate of 3.84 Mcps.
   b. cdma2000 (MC) uses a chip rate of 3.68 Mcps.

2. Pilot structure:

   a. Common pilot uses CDM.
   b. Dedicated pilot uses TDM.

3. Sync vs. async:

    a. WCDMA (DS) uses async/sync.

    b. cdma2000 (MC) uses sync.

The two proposed systems, WCDMA and cdma2000, will have common interface elements on the network side. Therefore we can call them two modes under a one G3G standard.

7.2.4 TORONTO MEETING

The following was decided at the Toronto's OHG meeting[1] and approved by the ITU.[2]

PHYSICAL LAYER (L1). The proposed common channel structure, with the position bits in accordance with WCDMA requirements, is defined, where the common pilot channel (CPICH), sync channel (SCH), primary common control pilot channel (PCCPCH), and secondary common control pilot channel (SCCPCH) have been defined.

    The dedicated pilot channel has a rate of 1/3 FEC code and a spreading factor of 256; the number of pilot bits applied to the dedicated traffic channel is under consideration.

HARMONIZATION REQUIREMENTS. The requirements for harmonization are listed below:

1. ANSI 41- and GSM MAP-based services should be fully supported in the Radio Access Network and associated with all three 3G CDMA modes.

2. Support functionality based on synchronous operation such as location calculation, and so forth.

3. Supports seamless handoff between the harmonized DS and MC, including IS-95 for ANSI 41 and the equivalent of this for UMTS/GSM.

4. Minimize the complexity of dual-mode and multiband terminals and equipment.

HARMONIZATION APPROACH. A conceptual diagram of the harmonization required to achieve these requirements for the DS and MC modes is shown in Figure 7.1. Note that this figure includes potential changes to the Physical Layer, L1.

The harmonization approach shown in Figure 7.1 has the following components:

1. For the DS mode the baseline starting point for supporting both core networks is

   a. L1, as stated previously
   b. W-CDMA L2
   c. W-CDMA L3 Radio Resource Control (RRC)

2. For the MC mode, the baseline starting point for supporting both core networks is

   a. L1, as stated previously
   b. cdma2000 L2
   c. cdma2000 L3 Radio Resource Control (RRC)

3. For the TDD mode, the baseline starting point for supporting both core networks is

   a. TDD Mode L1, the Physical Layer
   b. TDD Mode L2 as per 3GPP
   c. TDD L3 Radio Resource Control (RRC) as per 3GPP

4. The concept of hooks, as shown in Fig. 4.10, is defined as any functionality that is specified for the initial release of the standards so that the extensions that need to satisfy the requirements stated above can be defined in detail.

5. The concept of extensions, as shown in Fig. 4.10, is defined as any additional functionality at any layer that needs to be specified in detail to meet the requirements stated above, assuming the appropriate hooks are in place to enable the extensions to be defined without major changes to the baseline protocols.

BREAKDOWN IN PHASES.[2] Protocol Layers 2 and 3 for Direct Spread, Multicarrier, and TDD will be developed in two phases, as outlined below (including any consequential impacts on the physical layers):

Phase 1. The baseline parameters in all three radio layers, including the hooks as defined above, will be completed first.

Phase 2.   Completion of all the detailed specification of all
   the extensions to Phase 1 necessary to fully support ANSI
   41 and GSM core networks.

A more detailed view of the protocol architecture for the DS
mode connected to an ANSI 41 network based on the principles of Fig. 7.1 is shown in Fig. 7.2.

   3G operators may select combinations of the above protocol stacks subject to the requirements of their nation or
regions. The global specification for G3G must be sufficiently
detailed to enable operators to flexibly choose between the
various harmonized radio access and core networks.

7.3   A SIMPLE WAY TO APPROACH G3G DREAM[3]

This section will discuss reaching the G3G dream of the wireless mobile industry. Of course, we should ask ourselves first,
can a G3G standard be achieved easily?

7.3.1   CAN WE HAVE A G3G STANDARD?

Using the 3G standard can help achieve the goal of a future
phone handset that can be used anywhere in the world. Today,
GSM already benefits from this goal. But, GSM is a TDMA
system and is not a high-speed data system. For a high-speed
data system, CDMA is the right choice. Although not easy to
accomplish politically, the G3G standard we would like to
have is a single, universal standard CDMA system that can
technically be agreed on among all the international manufacturing companies.

   In 1998, 13 G3G proposed standards, based on a CDMA
system (see Fig. 4.3), were submitted to the ITU. The ITU's
responsibility is to find one standard among them; however,
this is impossible. Although the operators feel that the "convergence" approach, converging from 13 to one standard, is
the proper way to go, not many vendors support this.
Therefore, they are looking at the harmonization approach,
harmonizing from 13 to fewer standards, as an alternative.

The operators may try to have a handset that can be a multi-mode and multiband phone. This approach is not easy either. One idea is to develop software-defined radios (see Sec. 7.4). If this type of radio can be successfully developed, we do not need a 3G standard. The programmable function in the handset could be changed from one system to another simply by pushing a button. But, this cannot be accomplished in the near future.

In the meantime we can think about standardizing a universal smart card that can be inserted into any phone. For example, any different regional phone with someone's smart card inserted in it becomes an individual's personal phone. If this method is used, the roaming issue is solved and the G3G system is not urgent. The regional system standard can still be preserved, and the regional phones, with the universal smart card slot on them, can be rented at regional airports and harbors.

7.3.2   A SIMPLE SMART CARD CAN BRING G3G TO AN EXCELLENT 3G WIRELESS COMMUNICATION SYSTEM

If the universal smart card (standard card) is developed (see Fig. 7.4), a G3G standard is no longer an urgent necessity. We can have a regional 3G standard first. Each region can devote its effort to the development of a better system. No compromise is needed in developing a regional 3G system. For example, the dispute regarding the 4.096-, 3.86-, and 3.68-Mbps chip rates for the G3G standard no longer exists. The different chip rates have no impact on this simple solution. Furthermore, the advanced smart antenna (see Sec. 7.5) integrated with the base station can gain a high capacity. The IP-based network, with either ATM centric or router centric (see Sec. 8.2), can lower the cost of operation and increase the data rate. Of course, the voice quality and the performance of the system are the key factors that please the users and need to be well considered.

After the regional 3G systems are deployed, the users can evaluate them and then select the best one. The remainder of the regional markets could then, in time, adapt to the best regional system. Eventually, a universal 3G system would evolve.

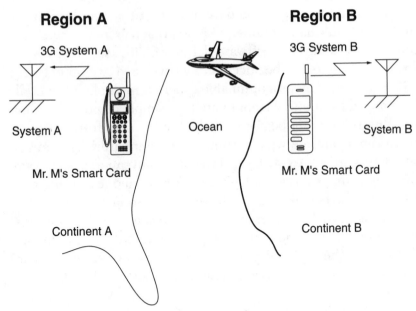

Figure 7.4.   Concept of a universal smart card.

7.4   SOFTWARE RADIO[4]

A software radio is a radio that can operate on any mobile radio system through software programming. The first idea of the software radio came from Don Steibrecher.* He called it the programmable radio in 1993. W. Lee in a meeting at Pactel suggested it be named the software radio instead of the programmable radio.

The software radio is still in the infant stage. The key element in this radio is its high-speed analog-to-digital (A/D) conversion. Today the digitizing can be 100 Mbps. Each sample can have 2^{12} levels.

Now the RF signal frequency has to be lowered and converted to IF band to around 100 MHz, such that A/D conversion can be used to digitize the signal into digital format. The digital signal processor (DSP) also has to have high speed to handle functions such as filtering, modulation, demodulation,

---

*Dr. J. Mitola has also been recognized as the first to publish the concept of software radio in 1992.

equalization, rake receiver, correlator, channel coding, encryption, and diversity. The ideal software radio will have the A/D converter placed before the high-power-amplifiers (HPAs) while transmitting and right after the LNA while receiving, as shown in Fig. 7.5. These two kinds of amplifiers need a broadband and linear device, which has been gradually achieved through several breakthrough technologies (see Sec. 7.17).

If the software radio technology can be ready soon, we may not need G3G. It can have an analogy that the G3G is a universal language that everyone has to learn how to speak. Software radio is like a person who can speak many languages and can talk to anybody using the local language. In this case the local base station is the local language. Each base station then can have its different 3G system. Only the handset has a software radio, which can change through software, to the local base station's system. With this changing capability, global roaming is achieved.

Alternatively, the handset can be from a different 3G system, and each base station can have a software radio that can analyze the received signal and pick the right system to use in response. In this case, each handset needs to preregister with the local base station; otherwise land-line-initiated calls will

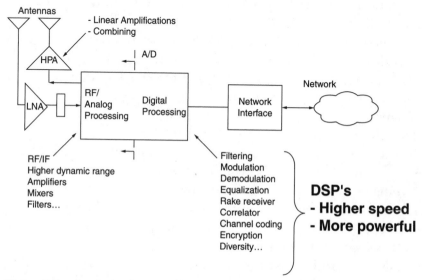

Figure 7.5. An ideal software radio.

have a hard time reaching the handset. Later, the name software radio is used for the ultimate device, and the name software-defined radio is used for the signal processing in the DSP and hardware support.

7.5   HOW SMART IS A SMART ANTENNA[5]

Smart antenna is a generic name the industry is commonly using even before any tests show if the antenna is smart or not. Some smart antennas may have a low I.Q. score. The smart antenna can be divided into two catalogues: blind detection and beam management.

Blind detection is a technique that uses antenna elements as smart sensors to detect two or more signals while receiving. When enough sensors are spread at the receiving end, the information from every incoming signal represented has a different signal intensity level at each sensor. Blind detection takes a matrix of information and goes through a deconvolution process using the relative sensor's location. The outcome can then decorrelate the individual signals. It can enchance the desired signal and suppress the undesired signal. The space diversity of an N branch antenna diversity is a subset of blind detection. Although space diversity only enhances a desired signal, it may be considered as a single-signal blind detection.

Beam management is divided into two techniques, beam switching and beam forming. Sometimes, beam forming is also called adaptive beam forming. Both use antenna arrays.

Beam forming techniques are relatively hard to apply to moving terminals with the required accuracy and latency. Therefore, beam switching is used in the present application. Beam switching is applied at the base stations for both transmitting and receiving. It switches the multiple antenna beams intelligently (see Fig. 7.6) following the mobile terminal moving from one beam to another. In this case, the interference is reduced. As shown in Sec. 3.5, reducing the interference means increasing the capacity.

Beam forming uses an adaptive antenna array, which is used in the military for antijamming on aircrafts. When beam

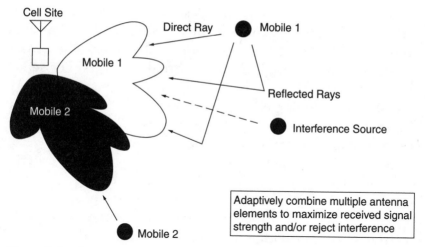

Figure 7.6. Smart antennas.

forming is used in the cellular system, the cellular environment is different from the aerospace environment:

1. It is an out-of-sight condition.
2. The receiver receives interference as well as multipath signals.
3. When the mobile terminal moves in the environment, the receiving signal strength is changing continuously. It is a nonstationary environment.

At present, beam forming is usually applied on WLL systems. In the future when the new DSP is faster, and the new algorithm is simpler, the convergent time for beam forming (including beam nulling and beam tracking) will be within the specified time range for cellular system use.

7.6 PERFORMANCE CRITERIA FOR THE VOICE AND DATA IN WIRELESS COMMUNICATION ARE DIFFERENT

Now the data communication era starts. For data communications we need to use a packet data format. Each packet contains a header that tells the destination address. Because of the packet data, a signal channel can be used to send different signals to different destinations. Therefore one physical channel can be used for many virtual channels. For sending the packet data we have to have packet switches. At present, the switching equipment is primarily circuit switches for serving voice. As soon as the packet switches are in place, the performance criteria for package data changes.

In the circuit switches for voice, the performance criteria are the block call rate, the dropped call rate, and the voice quality. In the packet switches for data, the performance criteria are the latency, the data throughput, and the real-time inquiry. If spectrum sharing applies to both voice and data, because of the difference in performance criteria, new guidelines have to be written and shared with the FCC. Because voice is a real-time operation, no delay can be tolerated, and the voice quality should be kept at an accepted level. On the other hand, the data is a non-real-time operation. Data errors can be traded off with a time delay for data correction retransmission.

The other main advantage of packet data is that the data link is virtually connected all the time. Therefore no dropped calls can occur. Because the voice is in data form in digital systems, packet switches can also transmit voice. But the following difficulties are bound to occur:

1. The voice packets over packet switches need to have priority.
2. When the voice packet is in the handoff region, the packet headers have to handle the location information.

Therefore, the voice over packet is an ongoing technology. The preferred packet protocol at present is the IP packet. The voice over IP (VoIP) today has been used on the wire-line Internet network with excellent performance. But the performance of VoIP in the wireless IP core network still needs to be investigated.

The GSM's packet switch is GPRS. Now it can only handle data traffic. Eventually, it will handle both voice and data services. G3G also is planning to have a wireless IP core network and handle both voice and data services in the future.

7.7 ATM SWITCH FOR PACKET DATA[6]

Today one packet switch is an asynchronous transfer mode (ATM) type of switch. The ATM switches use a "cell" structure, which should not be confused with cellular "cell." The ATM standard, or broadband integrated services digital network (ISDN) defines a cell as having a fixed length of 53 bytes, consisting of a header of 5 bytes and a payload of 48 bytes. Each cell's header contains a virtual channel identifier (VCI) to identify the virtual connection to which the cell belongs. The ATM technology, because of its flexibility and its support of multimedia traffic, draws much interest and attention. In wire-line and wireless communications, we need broadband switches, and ATM switches are the ideal ones for these applications. It was first developed for the wide area networking (WAN) equipment. ATM is designed to support multimedia traffic. It offers the benefit of handling the broadband signal channels needed for the increasing volume of data communications traffic. The ATM network is shown in Fig. 7.7. As mentioned previously ATM is a high-speed packet-switching technique that uses short fixed-length packets. The fixed-length cells simplify the design of an ATM switch at high-switching speed conditions. The standardized short-length cell reduces the time delay and the variance of time delay, which is jitter for delay-sensitive services such as voice or video. Therefore, short fixed cells are capable of supporting a wide range of various traffic types, such as voice, video, and image. An ATM switch will handle a minimum of several hundred thousand cells per second at every switch port. Each switch port will support a throughput

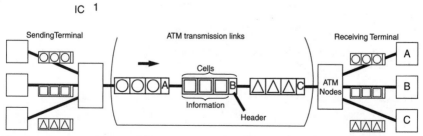

Figure 7.7. The ATM network.

of at least 50 Mbps, and 150 and 600 Mbps are proposed as standard ports. The ATM switch normally has 50 ports; more than 100 ports is considered a large switch. The general structure of an ATM switch is shown in Fig. 7.8. In an ATM switch, cell arrivals are not scheduled.

A number of cells from different input ports may simultaneously request the same output port. This is called output contention. A single output port can transmit only one cell at a time. Thus, only one cell can be accepted for transmission and others simultaneously requesting the port must either be buffered or discarded. The most significant aspects of the ATM switch design are

1. The topology of the switch fabric, such as time division and space division

2. The location of the cell buffers, such as internal buffering (sharing of buffers to reduce the number of cell buffers) and external buffering (for supporting the levels of priority for different classes of traffic)

3. Contention resolution mechanism [backpressure (sending back), deflection (routing) or loss]

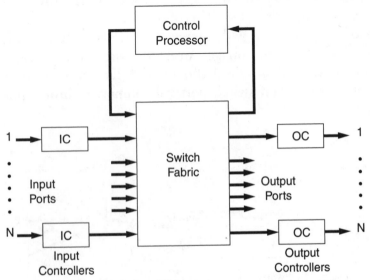

Figure 7.8.   Structure of an ATM switch.

ATM switches are connection-oriented, but when a connectionless server (a packet switch such as a router) is attached to an ATM switch, it can provide connectionless service (see Sec. 8.13). ATM can also use IP layer cells called ATM/IP to form an ATM-based IP network. In the future the ATM-based IP network will compete with the router-based IP network for low cost and better performance solutions. There are debates over these two. It is described in Sec. 8.7.

7.8   GPS-BASED SERVICES[7]

GPS is a U.S. system for finding the location either on earth or in the air. It is a MEO system, and its altitude is 11,000 miles. The rotation time of each GPS satellite around the earth is 12 hours. Because the location of each satellite (x_i, y_i, z_i, t_i) in space and its time t_i are known, where $i = 1,2,3,4$, we can find the ground unit location (a, b, c, t).

$$(a - x_1)^2 + (b - y_1)^2 + (c - z_1)^2 = (t_0 - t_1)^2 \qquad [7.1]$$

$$(a - x_1)^2 + (b - y_1)^2 + (c - z_2)^2 = (t_0 - t_1)^2 \qquad [7.2]$$

$$(a - x_3)^2 + (b - y_3)^2 + (c - z_3)^2 = (t_0 - t_3)^2 \qquad [7.3]$$

$$(a - x_4)^2 + (b - y_4)^2 + (c - z_4)^2 = (t_0 - t_4)^2 \qquad [7.4]$$

Therefore GPS can provide both the location on the ground and the elevation. From the above equations, if we make the elevation a known number or at ground level, c = constant. Then, there will be only three unknowns, a, b, t_0. This means we only need three satellites to obtain the ground location. If only two satellites are seen, the accuracy of the location becomes poor.

The GPS system needs 18 satellites to cover the entire earth. There are 24 satellites in space. Six of them are spares.

The GPS satellite system is a one-way transmission system. The satellites transmit the spread spectrum signals with a c-code (coarse) and a p-code (precision). The c-code is for commercial use and the p-code is for the government. The GPS receiver is placed on a moving vehicle or nomadic platform. Tremble Navigation had made a commercial GPS at a cost of about $5000 in 1985. Today a GPS receiver is a very inexpensive device and its cost can be under $100.

The ground location obtained from GPS is about 10 m 80 percent of the time, in an open area or outside buildings. On the street, between two high-rise buildings, the chance of observing three satellites is less, and the accuracy drops.

The normal GPS receiver cannot receive signals inside a building from the GPS satellite. But the highly sensitive GPS receiver is based on advanced hardware and software (DSP) and can have a 20-dB gain above the normal one. Only then can the location in the building be obtained. Snap Trak was the first company to demonstrate the in-building location.[8]

The GPS's service can locate airplanes, fleet, cars, and persons. Location-based services using GPS can be standalone services. A person can find his or her location as long as the GPS receiver has a software translation from the longitude and latitude values to the road name or address. Use of the GPS system is free; only the GPS receiver itself needs to be purchased. However, for the cellular/PCS system providers to use GPS location information, they have to find a means to receive the location information from the mobile unit and send it back to cell sites.

Another advantage of using the GSP system is to synchronize the wireless communication system for FDD or TDD modes. The GPS provides a master clock. It is free and accurate. Therefore, communications systems no longer need an asynchronized system.

But for security reasons some foreign countries may not like to rely on the United States' GPS master clock. These concerns should be alleviated because of the heavy use of GPS in many different services in the United States, especially in the E911 services. GPS should be a very secure source.

7.9   LOCATION TECHNIQUES AND SYSTEM ARCHITECTURE FOR E911[9]

The FCC has issued a mandate requesting that the cellular/PCS system providers carry out the E911 in two phases:

Phase 1, E911 report. Oct. 1998 (to serve any cell phones authorized or unauthorized)

Phase 2. E911 location. Oct. 2001

In Phase 2, the location accuracy is within 125 m at a probability of 67 percent. The response time for the location is not specified. It may have to be within the time of E911 connection. The 67 percent is also ambiguous. It could mean 67 percent of the total area, 67 percent of successful connections at any location, or 67 percent of handsets. The FCC has not clearly defined this yet.

7.9.1   LOCATION TECHNIQUES

There are three different basic techniques to find the location.

USING MOBILE-BASED LOCATION TECHNIQUES:

1. *Dead Reckoning.*   The mobile unit is equipped with a compass and a speedometer to track the location of the mobile. The cost is high and the accuracy cannot be controlled.

2. *Loran-C system.*   This is a location system used by the U.S. Coast Guard. It has numerous beacons located near the coast sending a 500-KHz-tone signal with their location information. The Loran-C receiver takes three or more beacon signals to triangulate and find the receiver's location. The location accuracy sometimes is very poor because of the number of beacons and the beacons' locations relative to the Loran-C receiver. In some areas, Loran-C beacons are not available, especially in inland areas.

3. *GPS system* is a very accurate system (see Sec. 7.8).

USING NETWORK-BASED LOCATION TECHNIQUES.

Receivers are located at the cell sites. Those receivers can receive the signal of the mobile unit from its (1) field strength, (2) angle of arrival, and (3) time of arrival (TOA) or differential time of arrival (DTOA).

Some receivers use one of the above techniques, and some use two or even three to increase accuracy. Then, based on the information about triangulation of three or more cell sites, the location of the mobile unit can be obtained. These network-based location techniques have some advantages and disadvantages:

1. All the handsets do not need to be modified.

2. The cost can be lower if the FCC's requirements do not need to be met.

3. The technique is simpler.

4. No need to have the location information transmitted from the mobile to the cell sites. The cell sites obtain the mobile location by themselves.

5. One problem of using this technique is the accuracy of the location. It is very hard to meet the FCC's requirement for location outcome. Also the uncertainty due to the multipath fading condition of each cell site makes this technique hard to pursue.

6. In cdmaOne systems, because of the power control algorithm, the mobile transmitter has to lower its power while approaching its own cell site. It means that the neighboring cell sites will receive weaker or unusable signals. The techniques then cannot be applied.

HYBRID SYSTEM.   This is a means of combining the techniques in mobile- and network-based systems to serve both the existing mobile units (or handsets) and the future mobile units (equipped with GPS receiver).

QUALCOMM'S SYSTEM.   This system utilizes the GPS receiver of each cell site in a cdmaOne system to be the known source. The mobile station receives a CDMA pilot signal from at least one cell site and acquires the system time from the pilot signal. The propagation delay between the mobile and cell site can be used to adjust the mobile system time to correspond to "true" GPS time. Then the time t_0 in Eqs. [7.1] to [7.4] is known, and we only need three satellites to obtain the location (a, b, c, t_0), or we need two satellites if the elevation parameter c is set as a constant. We can also use more known parameter values (from ground cell sites) to obtain a more accurate solution. Although the ground signal propagation is not accurate for time arrival calculation, it is much stronger for inbuilding reception.

7.9.2 E911 SYSTEM ARCHITECTURE

THE WIRE-LINE E911 SYSTEM ARCHITECTURE. The E911 call goes through local exchange telephone company (LEC), through a message line called the centralized automatic message accounting (CAMA) with the calling party number and the automatic number identification (ANI), and then to E911 tandem to E911 the central processing element (CPE). During this time, the features are critical in an emergency situation (see Fig. 7.9).

The calling party location is requested from E911 automatic location identification (ALI) to provide the subscriber location data (name, street address, etc.) from the ALI database to the router. This information will be provided to both the customer and the public safety answering position (PSAP).

Then the selective router automatically provides routing to the correct primary PSAP. The selective router can also auto-

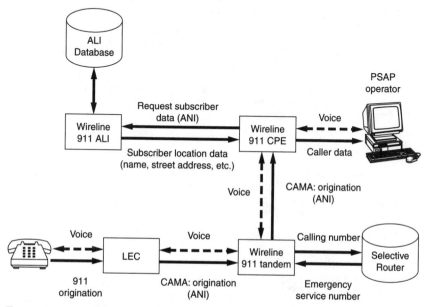

Figure 7.9. Architecture of the wire-line E911 system.

matically reroute the call to another PSAP, such as from police department to fire department.

THE WIRELESS E911 SYSTEM ARCHITECTURE.   In the wireless system (see Fig. 7.10) the E911 call goes through MSC, with the location information to the home location register/service control point (HLR/SCP). In the wireless system, the mobile subscriber does not have ANI because mobile units do not have fixed lines. Thus, ANI does not have any meaning, and ANI becomes a pseudo-ANI. The location information, the called party number, and so forth, are sent to wire-line 911 ALI. The voice path travels through MSC to wire-line E911 tandem to wire-line E911 CPE to PSAP. There will be many additional modifications to the system shown in Fig. 7.10 in the future.

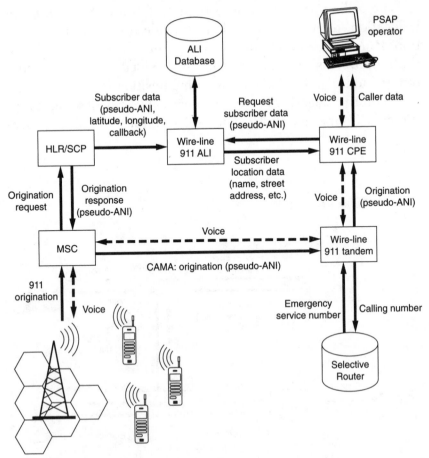

Figure 7.10.   Architecture of a wireless E911 system.

7.10 COMPUTER TELEPHONY (CT)

Computer telephony (CT) is an industry that concerns itself with applying computer intelligence to telecommunication devices, especially switches and phones. Since the CT will apply to the wireless phone network in the future, we have to understand its operation in the wireline. The computer can be a terminal or a replacement of a traditional switch. It takes advantage of rapidly improving PC/UNIX processing power, the programmable switch, and local area network (LAN) and IP technologies. The applications can

1. Integrate PC with switching cards through PCI bus, as shown in Fig. 7.11a.
2. Integrate PC with a programmable switch through transmission control protocol/Internet protocol (TCP/IP), as shown in Fig. 7.11b.

These arrangements create an open standard for fast applications and the programmable switch. CT technology is playing a key role in value-added services and switching-intensive products. Therefore CT can lower the network infrastructure costs.

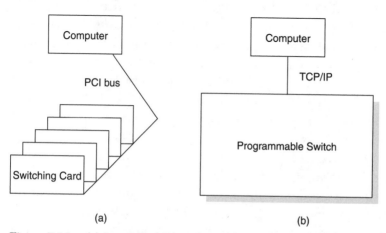

Computer

PCI bus

Switching Card

(a)

Computer

TCP/IP

Programmable Switch

(b)

Figure 7.11. (a) Integrating PC with switching cards through PCI bus. (b) Integrating PC with programmable switch through TCP/IP protocol.

7.10.1   NETWORK AND PROGRAMMABLE SWITCH

The CT application can meet new network requirements:

1. Wireless intelligence network
2. Interworking between networks
3. Mobile switching
4. Deploying faster on a variety of new services

The programmable switch is a general-purpose computer. CT is developing the programmable switch for the following reasons:

1. Expendability, scalability, and up-to-date architecture/ technology.
2. Can grow with computer technology.
3. The hardware and software can be less costly yet more powerful.
4. The open standard creates multivendor support for best pricing and services.

7.10.2   VARIETY OF APPLICATION

The current CT marketplace is complex and broad based. It includes a wide variety of application-specific devices like Interactive Voice Response (IVR) systems, voice mail systems, e-mail voice gateways, fax servers, switches (PBX and CO), automatic call distributors, and predictive dialers. The complexity of CT's marketplace is compounded by the need for a user device to integrate these devices with host-based computer systems, client-server systems, or desktop computer systems.

Services are being enhanced and new services are using emerging technologies like voice compression and expansion, data/fax/voice modems, desktop telephony interfaces, hearing impaired devices, and screen-based phones. With CT added, enterprise telephony networks have become too complicated to understand, manage, or perform as expected, and the duplication increases the cost of implementation and service. Figure 7.12 shows a possible network.

7.10.3 INTERCONNECTION AND INTEROPERABILITY

The complexity of CT has created huge interconnection and interoperability problems, dramatically slowing down the growth of the CT market. Software developers want to create integrated applications by combining features, services, and technologies as needed, regardless of supplier, vendor, technology, or industry source. Market forces have created a need for a set of agreements on the many interworking issues, agreements that allow the customer to select any hardware, platform, and application and put them together to build new services and agreements that allow users to enjoy straightforward implementations and avoid duplication of hardware, services, and administration.

Interoperability is crucial to market growth because it offers lower-cost products that are easier to install and main-

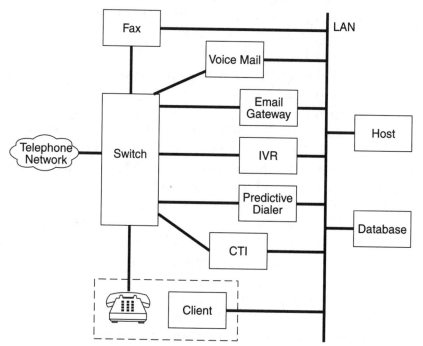

Figure 7.12. Architecture of a CT network.

tain, are faster to market, and provide more options for customers and suppliers. Thus, CT implementations must address five factors, modularity, extensibility, flexibility, determination, and resource sharing.

7.10.4   CT EVOLUTION

CT evolution can be divided into market and technology:

1. CT market evolution is from messaging applications (IVR, FAX, voice message network services, media server) to switching applications (call center, call completion, wireless switching, open switching integration, BCP).

2. CT technology evolution is from hardware applications (voice, FAX, PBX integration, multimedia API) to software applications (ATM, hot insertion, wireless interface, conferencing, dynamic resources, media, and switching API).

7.10.5   COMPUTER TO BE USED, UNIX OR PC

1. UNIX is dominating large CT systems. It is mature and stable and has higher processing power with a multiprocessor, multitasking, and fault tolerance. UNIX is a popular computer in the telecom industry.

2. PC (Windows NT) is focusing on small to medium systems. It starts picking up the speed on processor powers and increasing capacity on multitasking and redundancy. The multiprocessor Windows NT server is fast and costs less. A PC is easy and simple to install and has developed features. It may be the system for the future. CT has many high-potential low-risk emerging applications available. CT infrastructure developments focus on distributed architecture for low subscriber areas (50 to 250K), not for high-capacity systems. CT does have the evolutional path for multimedia core switch application.

Because the CT and wireless communication services will inevitably merge in the future, there are many issues that a CT application would need to be solved to meet the requirements of the wireless network.

7.11 INFRARED/MILLIMETER WAVE COMMUNICATION FOR HIGH-SPEED DATA

The infrared (IR) wave is an invisible light that can be generated by either laser or light-emit diode (LED) and received by the photo detector (PD). It has a higher propagation loss than electromagnetic (EM) waves. A high-power laser device could propagate close to 1 mile in 1998. But it could only be used when under line-of-sight conditions. Therefore we did not pay attention to this spectrum. Now the wireless information age has arrived. The high-speed data transmission (5 Mbps and up), with a high-capacity system to serve a large volume of traffic, needs a huge bandwidth. Only the millimeter (mm) or infrared waves can provide such huge bandwidth. Figure 7.13 shows the spectrum of IR and millimeter waves. The infrared application and the millimeter wave are beginning to draw attention. The millimeter wave can be generated by using either a Gunn or Impatt diode and can be received by the PIN diode.

The millimeter wave also has the properties of reflection and diffraction, which can be used to receive signals in the multipath environment, but the IR wave does not. Instead the

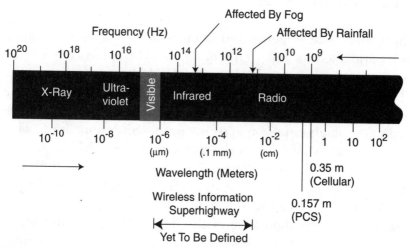

Figure 7.13.   The IR and millimeter spectrum.

IR transmission has to artificially create multipaths by using diffusion transmission technology. The diffusion transmission can be used in the out-of-sight condition.

The millimeter wave has great penetration through the fog but a great attenuation through rainfall. The IR wave is just the opposite, as it can penetrate rainfall but not fog, as described in Sec. 5.12. Because both waves have large bandwidths, we can design a dual-medium diversity receiver and receive the same information on both waves. During the fog condition, the millimeter wave signal can be received and during the rainfall condition, the IR light wave can be received.

This IR/millimeter link can only be used within 100 m to have a high reliability. For this reason, in the future, for a very high-speed data transmission, the IR/millimeter link will be the wireless portion of a hybrid system. The other portion will be the optical fiber link of the wire-line system. However, in most cases, people only use the voice in a vehicle, not the high-speed data transmission. Because of this, the 2- to 3.5-GHz spectrum could be sufficient for the mobility environment. The 5- to 7-GHz spectrum can be used for the nomadic environment. IR and millimeter waves are more useful for superspeed data transmission.

7.12 TURBO CODE[10]

The turbo code is generated from a family of convolutional codes and used for noisy environments. The turbo code can be used for transmitting high-speed CDMA data because the CDMA channel is closer to the complex gaussian noise in the multipath fading environment. Turbo codes provide the performance of channel capacity near the Shannon limit. This code increases the forward error correction (FEC) performances by adding a modest increase in decoder complexity. The 2-dB coding gain above the convolutional code is found. The longer the interleaver frame size, the better the performance. However, the cause of the delay also becomes longer. So, the turbo code is suitable for non-real time transmission, such as data, but has no advantage for voice transmission. The

Turbo encoder is situated at the sending end and the Turbo decoder is situated at the receiving end.

7.12.1 TURBO ENCODER

The Turbo encoder consists of two systematical recursive convolutional codes, first and second constituent codes, running in parallel. An interleaver precedes at the second recursive convolutional encoder, as shown in Fig. 7.14a. The mappings of interleaving for 3G two modes, WCDM and cdma2000, are different.

7.12.2 TURBO DECODER

This first decoder receives the systematic bits and parity bits from the first constituent code, as shown in Fig. 7.14b. The second decoder receives the parity bits from the second constituent and improves the performance on the soft-decision likelihood values. The process can be iterated many times, as shown in the returning path in Fig. 7.14b. The iteration process improves the performance, but the trade-off is in the delayed response.

Turbo code is accepted to be used in the G3G standard because high-speed data are a requirement of the wireless information age and the turbo code can meet this requirement to a great extent.

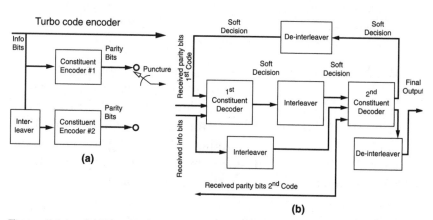

Figure 7.14. (a) Basic Turbo code encoder. (b) Basic Turbo code decoder.

7.13   CAN WDM BE USED IN MOBILE RADIO?

Wavelength division multiplex (WDM) cannot be used in mobile radio. Here we first explain the theory of WDM. The silica optical fiber experiences low loss in the wide wavelength range of 0.8 to 1.6 μm. Scientists have found efficient utilization in this low-loss, but huge wavelength band, transmission pipe. Multiple optical signals can be sent on different wavelengths through a single optical fiber. WDM transmission has two indispensable components: multiplexer and demultiplexer.

1. The wavelength division multiplexer at the transmitting end multiplexes several signals with different wavelengths onto a single optical fiber.
2. The wavelength division demultiplexer performs the reverse function. There are three popular WDM techniques, as shown in Fig. 7.15:

   a. Using a short-wavelength (high-pass) filter and a long-wavelength (low-pass) filter to separate light into its respective wavelength components.
   b. Receiving and passing parallel beams of light through a defraction and then grating the defracted lights after the grating light go in different directions. It corresponds to the separate beams with different wavelength signals.
   c. Passing the light through a prism. The angle of refraction varies according to the different wavelengths, which enables demultiplexing.

WDM devices are very small, less than one-quarter of a penny. WDM technology cannot work on the frequency spectrum for mobile radio. The wavelength of mobile radio frequency (800 MHz to 3 GHz) is too long. The prism would be too large to install over the radio link, making it impractical. Furthermore, in the wireless communication environment, the severe propagation loss and the impractical large reflectors prohibit WDM technology.

With EM wave characteristics, we usually use different polarizations (waves) to carry different signals, called polari-

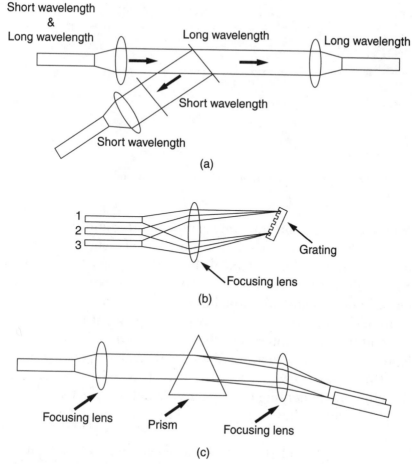

Figure 7.15. WDM device structure. (*a*) Filter type; (*b*) grating type; and (*c*) prism type.

zation division multiplexing (PDM). We can send eight signals on eight different polarizations, as shown in Fig. 7.16. Then we can say that the WDM technology used in optical fiber is analogous to PDM technology used in the wireless communication.

7.14   COMMENTS ON EQUALIZERS

The equalizer is used for canceling the multipath waves in the specified time delay spread intervals at the mobile receiver. Those multipath waves act like echoes. If all the echoes can

Transmitting Antenna   Receiving Antenna

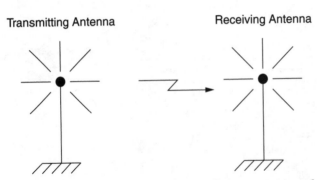

Figure 7.16. Using eight polarizations for transmitting and receiving.

be eliminated, the bit intervals can be shorter and the data bit rate can be higher. That is the purpose of using equalizers. But the equalizer has to cancel the incoming multipath waves at the right time. If the timing is missed, multipath waves cannot be canceled and additional noise is also added.

In addition, the equalizer may only implement three to four taps and handle an interval of three to four symbols. If the multipath waves of time arrival are longer than that specified interval, the delay spread will still interfere with the signal. Therefore implementing the equalizer is not a perfect solution.

Three approaches can be applied to eliminate the equalizer:

1. Use a high-level-state modulation such as 16, 32, or 64 QAM (quadrature amplitude modulation) to reduce the symbol rate but maintain a high bit rate. But the QAM is not a constant envelope modulation that can be used in a multipath fading environment (see Sec. 3.3). The QAM modulation can be used in a fixed wireless environment while the transceivers at both ends receive nonfading signals or when the mobile is close to the base station.

2. Use the directional antennas at one or both terminals to eliminate the multipath waves. In this implementation, the time delay spread can be reduced and the data transmission rate can be increased.

3. Use the diversity schemes at one or both terminals to smooth the signal fading. In the meantime, the time delay spread interval in the received diversity-combined signal is

reduced. For an M-branch diversity receiver, the time delay spread Δ_t is reduced by[11]

$$\Delta_t = \frac{\Delta}{M}$$

where Δ is the time delay spread of a simple fading signal. The transmission rate of an M-branch diversity receiver will be increased by M times the rate of a single branch receiver.

7.15 TRANSMIT DIVERSITY METHOD

The transmit diversity method is applied at the base station to increase capacity at the forward link.

7.15.1 TWO CATEGORIES—WITH AND WITHOUT FEEDBACK

1. Open-loop methods without feedback have three kinds of diversity schemes:

   a. *Orthogonal transmit diversity (OTD).* The same information with different encoded bits by adding the orthogonal properties is transmitted out of multiple antennas. It may be called Space-Time encoding (see Section 3.11.1). The OTD is the best among the three in the open-loop methods.

   b. *Time-switched transmit diversity (TSTD).* The same information will periodically (or randomly) switch to a transmit antenna.

   c. *Multicarrier transmit diversity (MCTD).* The same information will be sent by a subset of carriers and connected to spatially separated antennas.

2. Closed-loop methods with feedback have two kinds of diversity schemes:

   a. *Switched transmit diversity (STD).* The mobile receives the signal from the strongest base station antenna and feeds the information back to the base.

b. *Transmit antenna array (TAA).* The array weights are derived from the mobile measurement on the forward link and fed back to the base so that the multiple antennas transmit coherently.

7.15.2   ADVANTAGES OF USING THE TRANSMIT DIVERSITY METHOD

1. Combats fast (Rayleigh) fading on the forward link. Because the dual-antenna base-station diversity receiver has significantly improved reverse link capacity and ranges, the method matches up the capacity and range on the forward link.

2. Keeps the cost low and the size of handsets small and puts the transmit diversity at the base to achieve diversity gain at the mobile receiver. At the handset, the receiving diversity is hard to implement. First, the requirement of the antenna separation of half a wavelength cannot be implemented on the handset easily as the handset is getting smaller and smaller. Second, the cost of having a diversity receiver in the handset is high and makes the size of the handset bigger. Therefore it is not the desired solution.

2. Utilizes the existing dual-antenna at the base. The existing dual-antenna structure is used for the diversity receiver. It is very economical to use the same structure to apply the transmit diversity methods at the base.

7.15.3   CONCERNS OF USING THIS METHOD

1. The rule of thumb in wireless communication systems is not to transmit any signal unless it is necessary. In this case, if one single transmit power is split into two and fed to two transmit antennas, the power of each antenna is 3 dB less than the power before splitting. The receiving diversity does not give 3 dB less power to start with. If the two orthogonal encoded bit streams are transmitted, the 3-dB power loss can be regained by this additional orthogonal diversity gain.

2. The rake receiver is needed in the handset. In the TDD system, the transmit diversity at the base station will replace the receiving diversity at the handset completely. But in the FDD system, the transmit diversity at the base has to still go with the rake receiver in the handset.

3. The transmit diversity method has proven to be good in the nonshadow environment in which the diversity gain may not be appreciated. When in the shadow environment, the transmitting diversity provide the same performance as the receiving diversity. Because the handset has to have a rake receiver to perform the diversity anyway, we do not need the transmit diversity, which adds cost and provides no advantage.

7.16 WCS, LMDS, AND MMDS

7.16.1 WIRELESS COMMUNICATIONS SERVICES (WCS)

The 2.3-GHz spectrum has been allocated to the licensed WCS, as shown in Fig. 7.17. Bands A and B are the paired-bands operated in the 52 major economic areas (MEAs) shown in Fig. 7.18, and Bands C and D are unpaired bands operated in 12 regional economic areas (REAs), also shown in Fig. 7.18. WCS provides services in fixed, mobile, radio location or broadcast satellite communications and in WLL (wireless local

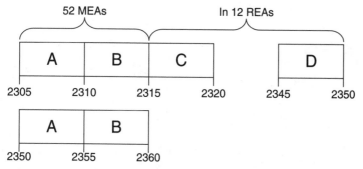

Figure 7.17. Wireless communications service (WCS).

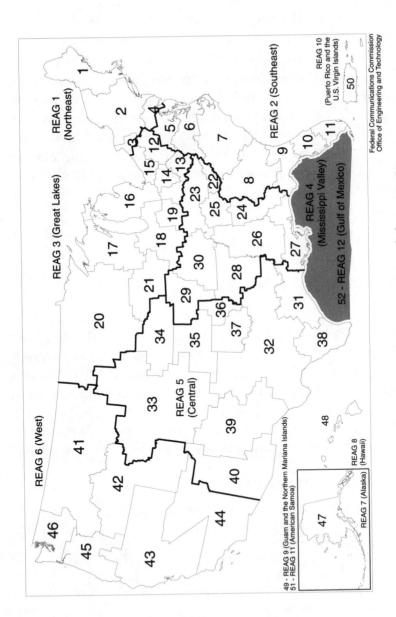

Figure 7.18. Regional economic area grouping (REAGs) and their constituent MEAs.

loop). The WCS spectrum is consistent with the international agreements concerning spectrum allocations. It is below the ISM band of 2.4 GHz. The advantage of using this spectrum band is that radio technology is mature in this spectrum range and the cost is very effective.

7.16.2   LOCAL MULTIPOINT DISTRIBUTION SYSTEM (LMDS)

LMDS operate at two spectrum blocks:

1. LMDS Block A

   a. 27.50–28.35 GHz (850-MHz bandwidth)
   b. 29.10–29.25 GHz (150-MHz bandwidth)

2. LMDS Block B

   a. 31.00–31.075 GHz (75-MHz bandwidth)
   b. 32.225–32.300 GHz (75-MHz bandwidth)

The spectrum of LMDS is around 30 GHz. Rainfall attenuates the signal in this spectrum range (see Sec. 5.13). Therefore it is used for a short distance (i.e., 100 m or less). Because the cost of the radio unit is relatively high in this spectrum range, vendors have to find a way to convince the operators or wireline companies that LMDS can be a cost-effective solution.

7.16.3   MULTIPOINT DISTRIBUTION SERVICE AND MULTICHANNEL MULTIPOINT DISTRIBUTION SERVICE (MDS AND MMDS)

MDS is a service that operates at 2.1 GHz. It has two 6-MHz wideband channels. The allocation of the spectrum from MDS, MMDS, and Instructional Television Fixed Service (ITFS) is shown in Fig. 7.19.

MDS and MMDS have their effective radiated power (ERP) limited at 2000 W for every 6-MHz channel. The total bandwidth for MDS is 12 MHz and for MMDS it is 66 MHz. The MMDS is used for the WLL and also can be used for two-way wireless Internet.

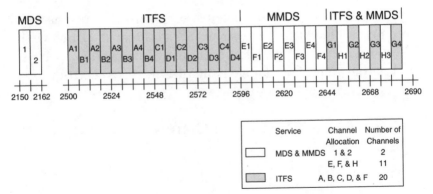

Figure 7.19.   Spectrum allocation of MDS, MMDS, and ITFS services.

7.17   BREAKTHROUGH IN BROADBAND POWER AMPLIFIERS

The broadband PA was always very difficult to achieve in the past because of the linearity of amplification. There are three different major approaches. One uses the distributed approach shown in Fig. 7.20a. All PA elements are either class A or class AB amplifiers, which are linear amplifiers. The linearity is used to prevent intermodulation (IM). The linearity between input and output in power dynamic range in Fig. 7.20a is extended because each element represents its power in a different range. The four radios share 12 PAs in a broad linear fashion, as shown in Fig. 7.20a. If one of 12 PAs is out of order, the impact on the performance of the whole amplification is very small. Another approach is shown in Fig. 7.20b. It uses the TDM approach to relax the linearity requirement. In using TDM, one radio is active at a time in its time interval. During that interval, no other radio signal shares the PA. Therefore no IM is generated. In this case, we can use a single class C amplifier with very high power. This second approach has been used by Fugent,[12] an entrepreneur company.

The third approach is applying the old technology, using a traveling wave tube (TWT) for terrestrial communication. TWT is a high-power, long-life highly efficient power amplifier that is being used on the satellites. Its advantages are

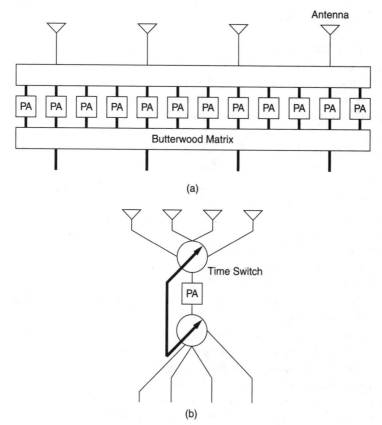

Figure 7.20. (*a*) Distributed arrangement. (*b*) Time-division arrangement.

1. It can generate high peak and average power. An ultimate goal of Hughes TNT is 2000 W peak and 200 W average. TWT is very efficient and has about 20 percent power output.

2. The linearization achieved by applying feed-forward techniques can have ±0.1 dB/50-MHz amplitude flatness and 40 to 50 dB gain in the frequency range of 1.5–2.6 GHz. The absolute delay is less than 10 ns. When the power output is less than 100 W, the efficiency of solid state device drops to 20 percent and that of TWT increases to 20 percent when the power output is greater than 100 W. However, the cooling needed and the size of the package make TWT less attractive at present.

7.18 REFERENCES

1. OHG "Harmonization Framework Agreement", June 3, 1999, Bell Mobility Co., Ottawa, Canada.

2. ITU-T "Recommendation Q. 1701 and its Supplement", Q. 170 (Framework for IMT-2000 Networks) has been identified a Q8/11, Rapporteur Meeting, Ottawa, September 8–17, 1999.

3. W. C. Y. Lee, "Can 3G Wireless Communications Systems Be Technically Excellent?" 1999 IMT 2000 3G Wireless Technology Conference, New Orleans, Louisiana, Feb. 10–12, 1999.

4. FCC Notice of Inquiry, "Inquiry Regarding Software Defined Radios," ET Docket No. 00-47, March 17, 2000.

5. W. C. Y. Lee, "How Smart is the Smart Antenna?" The Fifth Annual Workshop on Smart Antennas in Wireless Mobile Communications, Stanford University, July 23–24, 1998.

6. U. Black, *Foundation for Broadband Networks*, Prentice Hall, Englewood Cliffs, New Jersey, 1995.

7. E.D. Kaplan, *Understanding GPS: Principles and Applications*, Artech House, Boston, 1996.

8. Steve Poisner, "Review of the GPS-Based E911 Technology." The Location Implementation Conference for Phase Z E911 Location Technology, San Francisco, CA, August 26–27, 1998.

9. Revision of the Commission's Rules to Ensure Compatibility with Enhanced 911 Emergency Calling Systems, CC Docket No. 94-102, Report and Order and Further Notice of Proposed Rulemaking, 11 FCC Red 18676 (1996), 61 Fed. Reg. 40348, 40374 (1996) (*E911 First Report and Order*) (*E911 Second NPRM*); Memorandum Opinion and Order, 12 FCC Red 22665 (1997), 63 Fed. Reg. 2631 (1998).

10. Steven S. Pietrobon, "Implementation and Performance of a Turbo/Map Decoder," *International Journal of Satellite Communications*, pp. 4–17, 1998.

11. W. C. Y. Lee, *Mobile Communications Engineering Theory and Application*, Second Edition, McGraw-Hill, New York, 1993, pp. 391–392.

12. Fujant Product "Sampling Amplifier," Fujant, Carpenteria, CA.

INTERNET AND WIRELESS FUTURE

8.1   OVERVIEW OF THE INTERNET

8.1.1   History of the Internet

The Internet began in the late 1960s from Advanced Research Project Agency's ARPANET project to build a packet-switched network. In 1970, the ARPANET project grew to support the Department of Defense and other government and research organizations. The term *Internet* was first used in 1983 to describe the concept. In 1985, National Science Foundation (NSF) funded several supercomputer centers and used a 56-kbps ARPANET network, called NSFNET, to link them. NSF allowed any regional or university computer centers that could reach this NSFNET network by connecting it. This was the seed of the Internet as we know it today. In 1987, NSF awarded a contract to Merit-Network Inc., which is in partnership with MCI, IBM, and University of

Michigan, to upgrade and operate the NSFNET backbone by using 1.5-Mbps T1 leased lines to connect six regional networks, five existing supercomputer centers, and other sites such as university sites. On July 24, 1988, the old 56-kbps network was shut down. From 1989 to 1991 the Merit/IBM/MCI team proposed and upgraded to a higher-speed 45-Mbps backbone, which expanded to 16 regional sites and connected over 3500 networks. In 1993, NSF decided to get out of the backbone business and solicited bids for building Network Access Points (NAPs), from which the commercial backbone operators could interconnect. In 1994, four NAPs were built. In 1995 NSFNET was essentially shut down and the NAP architecture became the Internet.

8.1.2   Internet Architecture

The Internet is a packet-switched network. The packet-switch network has the following attributes.

1. There is no single, unbroken connection between the sender and the receiver.

2. The data stream is split into IP packets, and each packet contains its address. Therefore the data stream is sent out without establishing a dedicated connection at beginning and each packet is routed independently, possibly over a different route. Thus the packet-switched system is a connectionless system in contrast to a circuit-switched telephone system, which establishes a connection for each call from the sender to the receiver and dedicates network resources to it (see Section 7.7). We call the packet-switch network a connectionless network because it does not have a dedicated connection. Connectionless networks as of today cannot offer the quality of service (QoS), such as the latency and throughput.

3. The packets can be lost, duplicated, corrupted, and/or arrive out of order at the destination.

The Internet is made up of hardware, software, and communication links that provide users with a ubiquitous platform from which to run applications and access information. The fundamental entities in the Internet are clients, routers, and gateways, as shown in Fig. 8.1.

1. A client is generally an application on a user device, such as a computer that facilitates setting up either communication or information sessions with other users or applications. Netscape's Web browser is an example of a client.

2. A server is a combination of hardware and software that satisfies a client's request for information or communication. Yahoo's stock and travel quote services are examples of a server application.

3. A router is a device that routes traffic among different networks [e.g., from an enterprise network to an Internet Service Provider (ISP) network].

4. A gateway is an entity that generally transforms traffic from one type of network to another type (e.g., conventional telephone signals to IP packets for Internet telephony). The routing functionality usually handles multiple gateways.

5. Network Access Point (NAP) interconnects the commercial backbone operators.

8.1.3   TCP/IP

TCP/IP is the technology platform for the Internet. Internet has no single, unbroken connection between the sender and the receiver. The data is split into IP packets, each of which has an address overhead. Each packet is routed independently, possibly over different routes, which is in contrast to a circuit-switched telephone system, which establishes a specific connection and dedicates part of the network to each call. The IP networks are connectionless networks, and circuit-switched networks are connection networks. In the connectionless network, packets can be lost, duplicated, corrupted, and/or arrived out of order at the destination. There are two protocol-handling layers, Layers 3 and 4 (see Fig. 8.2).

THE INTERNET PROTOCOL.   The user's data is transformed into IP packets. IP enables routers along the way through the network to send the packets to the right destination.

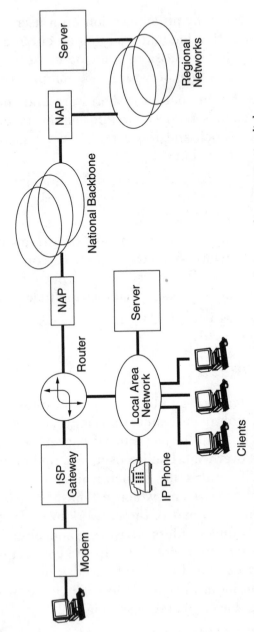

Figure 8.1. Structure of the Internet, comprised of hardware, software, and communication links.

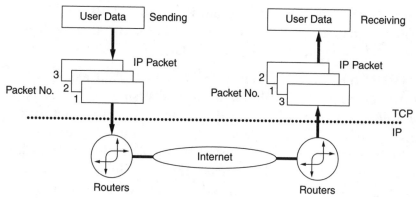

Figure 8.2. The protocol-handling layers.

It is a best-effort delivery, and packets may be lost along the way. The wide bandwidth (in the fiberoptics) of the network can reduce the loss of packets in the wire-line Internet network. This IP protocol uses Layer 3, the Network Layer.

THE TRANSMISSION CONTROL PROTOCOL (TCP). A header that includes a packet identifier, a checksum, and source and destination IP address is added to each of the IP packets. TCP is the Application Layer, or Layer 4. At the receiving end, the checksum of each set of data is calculated and the packet identifier tracked. The checksum is a computed value that depends intimately on characteristics of a particular set of data, and the checksum changes if the data changes. The packets will be asked for retransmission if packets are lost or corrupted. In this layer, once they are properly received the packets are reassembled into the original data format.

TCP CONNECTION. A network application often has to identify the connection endpoints that are receiving data through some connection over the network. If an application running on a network client needs to send a file to its remote server, the protocols invoked by the application need to be formatted to the data according to specifications. Every data transfer between network endpoints has to be tightly controlled by conformance to interoperable network protocols such as TCP, the User Datagram Protocol (UDP), or the Real-Time Transport Protocol (RTP). The latter two deal with real-time data such as voice or priority data.

ADDRESSING IN THE INTERNET. Addressing in an Internet network is different from addressing in the telecommunications network. Each device in an Internet network is given an IP address, which is a 32-bit number in the current IP implementation [i.e., Internet protocol version 4 (IPv4)]. A more abstract textual address, arranged in a hierarchy of domains, is commonly assigned to a device or a user with a name to keep track of the association between the IP addresses and the names (see Fig. 8.3).

The top-level domains in the United States are

com for commercial organization
gov for government organizations
edu for educational institutions
net for network (e.g., ISP network)

International domains are

au for Australia
uk for United Kingdom
in for India
us for United States

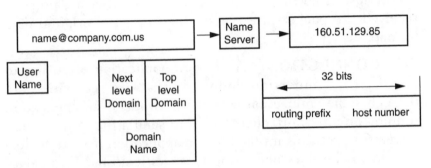

Figure 8.3. Addressing in an Internet network.

The next-level domain identifies a specific organization and a host computer within the higher-level domain. The host computer looks up the user name and delivers information to the user. All communicating entities on the Internet must have an IP address. This implies that mobile phones and terminals, when connected to the Internet, also must have IP addresses. For a device that is not permanently attached to the Internet (e.g., a dial-up mobile data device), an IP address is temporarily assigned from a pool of addresses for the duration of the connection.

8.1.4 SECURITY ON THE INTERNET

The security of access and transactions performed on the Internet becomes crucial as commercial use on the Internet increases. Corporate networks, banking services, stock services, and so forth, are generally protected from malicious attacks by the use of "firewalls." A firewall is a combination of hardware and software. A typical firewall, shown in Fig. 8.4, includes the following:

1. A secure server, which is the primary point of contact for connections from the Internet for service authentication and to protect the secrecy and integrity for transactions such as banking and credit card payments, receiving e-mail, and accessing a corporate database. It also processes any requests from the internal corporate network to the Internet, such as browsing the Web or downloading files. It can be used to log the Internet traffic between the internal

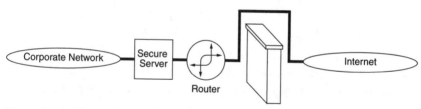

Figure 8.4.   The concept of a firewall.

corporate network and the Internet, and downloads to the level of every IP address accessed, data and time of access, number of bytes downloaded, and so forth.

2. A router examines packet headers and allows only certain types of packets to be sent or received and blocks other packets.

3. In addition to the primary point of security check, such as the secure server mentioned in item 1, secure electronic transaction protocols are used to authenticate and protect the secrecy and integrity of transactions such as banking and credit card payments on the Internet.

8.2   FUTURE OF IP NETWORKS

8.2.1   IP NETWORK STANDARDS

IP network standards are evolving to support the long-term growth, quality, billing, and mobility. The wireless Internet does not have enough bandwidth (air link) in the system. Therefore, the current Internet connectionless system, with packet transmission, can cause a long latency and packet loss if the system is used for the wireless mobile system. As a result, the IP network standard will need to be modified.

LONG-TERM GROWTH

1. IP addresses are scarce commodities. The rapid growth of the Internet is quickly exhausting the numerical address space. The new IP version 6 (IPv6) increases the number of bits for addressing from 32 to 128, accommodating a much larger address space. Besides, IPv6 also supports a graceful transition from IPv4, and the autoconfiguration of the network by having plug and play capability. The autoconfiguration of the network can automatically detect which devices and applications are attached and removed from the network. IPv6 is being promoted in the IP network industry.

2. Because of the fast growth of the Internet, today's Internet may not meet tomorrow's needs. Development of a second-

generation of the Internet, called Internet 2, has started. The Internet 2 development group consists of major universities and government agents. The goal of Internet 2 is to develop a high-volume, high-usage network.

3. The wireless Internet will become a hot trend. The wireless IP network will replace the current conventional network with limited bandwidth. However, many issues such as nomadic mobility and mobilized mobility, quality of service, and so forth, need to be solved (see Sec. 8.4).

QUALITY OF SERVICE.   A variety of QoS techniques are being constantly developed and enhanced to support real-time applications, such as voice over IP and video conferencing over the Internet. These techniques are used to reduce latency and maintain quality. They include (1) assigning different priorities to different types of traffic, (2) reserving bandwidth or processing capacity (a priority for certain types of connections), (3) supporting policy-oriented schemes for prioritizing different types of users or applications, and (4) improving the Internet protocol header format. A combination of these techniques, along with judicious network design, are already placing Internet telephony quality metrics within the guidelines specified by the ITU.

AUTHENTICATION, AUTHORIZED, ACCOUNTING (AAA) AND BILLING.   AAA and billing are needed because of the increased use of the Internet for commercial purposes. The Internet community is developing standards for AAA. Several commercially available routers are capable of collecting detailed accounting data for billing purposes. In the wireline Internet, AAA function takes care of the accounting but has no billing function. In the wireless Internet, the accounting function is very complicated and should be separated from AAA. In wireless mobile systems, the billing function needs to be continuously updated according to the marketing strategy demand. Therefore, the accounting and billing should be combined as one utility in the wireless mobile system.

MOBILITY.   The recent IP standard supports mobile devices accessing Internet applications. It is generally referred to as mobile IP capability. Although mobile IP is currently targeted to support packet data for nomadic mobility such as laptop computers, it will apply to mobilized mobility such as cellular data transmission. Also, it could conceivably be used to support mobile voice as well.

8.2.2   VOICE OVER IP (VoIP)

FIRST GENERATION OF VoIP.   The Internet is primarily used for data; VoIP originally was a concept of using the Internet for voice communications. In 1995, VocalTec Co. brought the VoIP into the Internet mainstream by introducing an Internet telephony client software package that runs on PCs. This first-generation VoIP product could be used only for PC to PC communications and was dismissed as a hobbyist toy for making free long distance telephone calls for the following reasons:

1. The users at each end needed a multimedia computer equipped with a sound card, microphone, and speakers.
2. The same software was needed on both ends.
3. Both users had to be on-line the moment the call was placed.
4. The conversation time was primarily spent tweaking controls for volume and compression to hear each other better.

SECOND GENERATION OF VoIP.   In 1996, IDT Corp. introduced its breakthrough Net2Phone services, allowing Internet users to place voice calls to any conventional telephone in the world. In 1997, IDT introduced Net2Phone Direct, allowing voice calls between two conventional telephone sets to be bridged via the Internet, eliminating the need for a multimedia computer and the client software. In 1998, a flood of new products supporting VoIP were introduced with focus on supporting the following:

1. Improved QoS
2. Carrier-class reliability
3. Scalable systems for carrier applications

ANATOMY OF A VoIP CALL—A THIRD GENERATION.
An origination gateway receives a call origination request and
dialed digits from a PBX or a central office, translates them
into an IP address of a destination gateway supporting the
called number, and exchanges information with it for call
setup. The origination gateway converts analog voice into a
digital signal through a digital signal processor (DSP). It does
voice encoding, header compression, priority queuing, and
routing. The destination gateway buffers, sequences, decom-
presses, decodes, and undigitizes the packet stream for delivery of voice to the called phone. The access gateways can also
be directly connected to conventional telephones, fax
machines, or IP phones (see Fig. 8.5).

QUALITY OF VoIP CALL.   The quality of a VoIP call
depends on two key factors—delay and delay variation. The
long delays impair conversation by (1) causing the listener to
start to talk before the sender is finished and (2) making echoes
(hearing oneself) noticeable and distracting. The delay variation
is also referred to as "jitter," it causes gaps in speech pattern
that results in jerky speech. In the current circuit-switched network, because of dedicated bandwidth for voice calls, the networks have predictable fixed delay and delay variation.

International IP voice calls on public ISP networks typically
have a 100- to 150-ms delay. This compares with a 30- to 50-ms
delay for terrestrial PSTN circuit networks and a 200- to 250-
ms delay for satellite circuit networks. The maximum tolerated
response delay is within 200 ms based on human behavior.

Improvements to VoIP are occurring very rapidly and are
as follows:

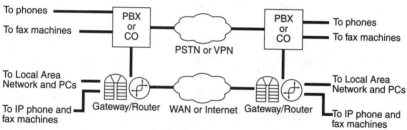

Figure 8.5.   The anatomy of a VoIP call.

1. In IP network, delay can be minimized by using one or more emerging IPQoS techniques.

2. Jitters are handled by dejitter buffers, where the received packets are stored before being delivered to the receiver. The received packets, then, reach the receiver in a constant delivery rate. The chosen size of the dejitter buffer determines the length of time the speech is stored prior to being passed on to the end user. Dejitter buffers smooth out delay variation but add delay, typically 50 ms, to the conversation.

3. An IP packet header containing 32 bytes each of source and destination address and several bytes of other information adds considerable overhead to speech packets. The solutions, such as Compound Real-Time Transport Protocol (CRTP), shorten the overall IP header to 4 or 5 bytes.

4. User Datagram Protocol (UDP) is used for VoIP. It does not require retransmission of lost or corrupted packets. Therefore, UDP is used instead of TCP, which requires retransmission in VoIP systems.

5. Most VoIP gateway products today support the wire-line environment. They assume an analog voice input and digitize and code the analog voice (speech signal) using one or more ITU vocoding standards. This process can be completely bypassed for mobile to mobile calls in cellular digital networks. For mobile-originated calls, the signal is already digitized and vocoded. VoIP gateways need to be aware of these variations and support mobile vocoding standards.

IMPEDIMENTS TO GROWTH. Voice over IP is poised for dramatic growth. The total voice traffic on IP networks in the United States is predicted to grow about 20 percent from 2000 to 2003. The ingredients to growth can be stated as follows:

1. Much of the growth today in VoIP on public network is driven by the arbitrage opportunity. ISP does not pay access charges for completing long distance calls, whereas the

interchange carriers do. As the Internet telephony grows, the regulatory environment may change to eliminate this opportunity.

2. Despite the widely used H.323 ITU standard for setting up voice calls over IP networks, interoperability between the voice gateway equipment of various vendors is still an issue due to propriety extensions and selection of different options for implementing calls. This is, however, improving with industry efforts to perform interoperability testing and provide smooth interworking of equipment.

3. The cost per single voice channel (DSO port) is still relatively high, perhaps due to early technology cycle and arbitrage value. This will have to come down rapidly, as is widely expected, for continued growth in VoIP traffic.

4. As more quality-conscious Internet telephony service providers (ITSPs) deploy VoIP on a larger scale, the reliability of the technology will have to be firmly demonstrated.

8.2.3 MULTIMEDIA AND MULTICASTING OVER THE INTERNET

AUDIO.   Interviewers, music, sound chips, and so forth, can be heard from radio stations through the Internet, but the old technology was limiting in that it had to download an entire audio file after it received it, prior to playing it. It was not uncommon to take 15 minutes to download a file that had less than 1 minute of sound on it. A newer use of audio on the Internet is called streaming audio, which allows the audio to be played without having to download the entire file first. Instead, one listens to the audio while it downloads onto the computer.

VIDEO.   Internet technology is moving beyond electronic mail and audio. Today, Internet technology has tools for video conferencing and collaborative "white board" applications. As with audio, the streaming video technology compresses video files dramatically and allows the receiver to start playing the video while the file is being transmitted.

MULTICASTING.   Usually the telecommunication networks that perform well at switching are not good at the multicasting or the broadcasting and vice versa. However, Internet technology supports both switching/routing and multicasting/broadcasting equally well. It has very powerful multicasting capabilities.

1. Multicast groups and sessions can be dynamically created and changed. A speech can be translated into different languages and multicast simultaneously.

2. Any user can join or leave the multicast group without the source having to do anything special and without the need for an intervening operator.

8.2.4   SCALABILITY OF IP NETWORK

The IP network elements have been historically designed to support enterprise applications. Now the technology can allow the IP network to scale up to support millions of users in carrier networks.

ROUTERS.   The capacity of routers has been increasing dramatically. The new Packet Over SONET (POS) systems were deployed with routing throughput up to 2.5 Gbps (OC-48) in 1998. The routers and fiber-optic networks with a throughput of 10 Gbps (OC-192) were deployed in 1999. The routers for wireless communications are developing by adding gateways of many necessary functions, starting in 2000. Also the router interface to the current BSC or BTS is another task related to the H.323 and IS-634 standards. This interface will connect the mobile radio access equipment to the Internet network.

GATEWAYS.   Prior to 1998, most gateways were enterprise gateways supporting several hundred to thousands of users in the corporate private network environment. The implementation of VoIP gateways can be different, ranging from dedicated standalone devices to add-on cards for switches, routers, or PCS. Scalability of these gateways, often as a limitation of the VoIP technology, is attempting to design the successive generations of VoIP gateways. In 1998, carrier-class gateways were capable of carrying 100,000 VoIP connections (ports). The further improvement in the design, cost, and capacity of carrier-class gateways is expected to support public telephony network infrastructure.

SERVERS.   The servers essentially determine the number of simultaneous transactions that can be supported per second. Parallel processing architecture and server farms (multiservers actively sharing the work load) are increasing the transaction processing capacities considerably.

8.3   WIRELESS LANS[4] AND IEEE 802.11[5]

8.3.1   INTRODUCTION

Standard LAN protocols, such as Ethernet, which operates at a fairly high speed with inexpensive connection hardware, can bring digital networking to almost any computer. Sharing of the information and distributed computing is beginning to be realized. However, the LAN is limited to the physical, hard-wired infrastructure of the building. Now, we would like to have wireless LAN to increase mobility and flexibility. An ad hoc network can be brought up and torn down or its configuration changed in a very short time as needed. Also, a wireless LAN is more economical to use. There is no need for expensive wiring or rewiring. But, the wireless LAN needs mobile IP. Mobile IP is used to attain wireless networks at the network layer, working with IP version 4 (IPv4). In this protocol, the IP address of the mobile machine does not change when it moves from a home network to a foreign one. The mobile agent, through mobile IP, tells a home agent on the home network to which foreign agent their packets should be forwarded. The mobile IP works only for IPv4 and does not take advantage of the features of the newer IPv6. Mobile IP is described in Sec. 8.4.

8.3.2   ARCHITECTURES

There are two different ways to configure a network, as an ad hoc network and as an infrastructure network. In the ad hoc network, computers are brought together to form a network "on the fly," as shown in Fig. 8.6. The infrastructure network uses fixed network access points with which mobile nodes can communicate (Fig. 8.7).

8.3.3   LAYERS

IEEE 802.11 standardizes the specifications on the parameters of both the Physical (PHY) Layer and the Medium Access Control (MAC) Layer of the network. The PHY Layer handles the transmission of data between nodes using either direct-sequence spread spectrum or infrared (IR) pulse position modulation for a data rate of 1 or 2 Mbps. For spread-spectrum transmission it uses the industrial, scientific, and medical (ISM) band, 2.4 to 2.485 GHz. For IR transmission it uses 300 to 42,800 GHz. The MAC Layer is responsible for maintaining order in the use of a shared medium. The 802.11 standard specifies a Carrier Sense Multiple Access with Collision Avoidance (CSMA/CA) protocol.

8.3.4   AVOIDING THE "HIDDEN NODE" PROBLEM

The hidden node problem, shown in Fig. 8.8, is that node A can communicate with node B, and node B can communicate

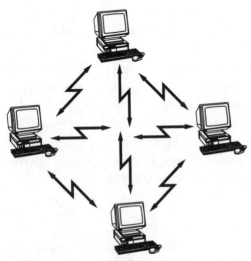

Figure 8.6.   An ad hoc network.

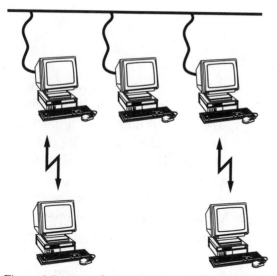

Figure 8.7. An infrastructure network.

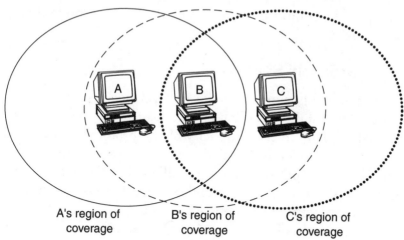

A's region of coverage

B's region of coverage

C's region of coverage

Figure 8.8. The hidden node problem.

with node C, but node A cannot communicate with node C. In this protocol, node A is alerted that node B is busy (with node C), and hence, it must wait before transmitting its packet to node C through node B. The 802.11 standard provides a reliable means of wireless data transfer, but it is not very suitable for the cellular systems.

8.4 MOBILE IP[6]

8.4.1 DEFINITION OF MOBILITY

Mobility and portability are similar terms. Currently, most mobile computer users are satisfied with portable operation. A computer can be operated at any group of points of connections but not during the time that the computer changes its point of attachment. The latter condition is mobility. Future mobile users want to have the network support uninterrupted connectivity between application and resource during the mode of operation.

Mobile IP is a protocol that allows truly mobile operation; by using the protocol, neither the system nor any of the applications running on the system need to be reinitialized or restarted, even when network connectivity is frequently broken and reestablished at new points of attachment.

The nomadic users of today's Internet are often satisfied with portable computing. Thus mobile IP does not provide much benefit.

8.4.2 MOBILITY SUPPORT

Providing mobility support at the IP layer needs great caution because the mobility problem can be transformed into a routing problem. Due to the simplicity of the protocol and related small amount of code needed, the necessary changes are implemented to the route table handling the home agent and foreign agent. Furthermore, mobility has other effects on protocols at every level of the network protocol stack.

For the needs of mobile voice connections and military applications, the Physical and Links Layer protocols in the mobile networking below the Network Layer have to handle adaptive error correction, data compression, data encryption, power minimization, and so forth.

8.4.3 TCP CONSIDERATIONS

TCP TIMERS. When high-delay or low-bandwidth links are in use, the default TCP timer values in many systems may cause retransmission or time outs, even when the link and the network are actually operating properly with greater than usual delays. The mobility-aware vendors will make more TCP timer values.

TCP CONGESTION MANAGEMENT. Mobile nodes are often used in wireless media, which are more likely to introduce errors, causing more packets to be dropped. This introduces a conflict with the mechanisms for congestion management in the modern version of TCP. Currently when a packet is dropped, the TCP implementation is to react as if there is network congestion. But the packet dropping is due to the mobile media at mobile nodes. A solution will be found in the future.

8.4.4 TERMINOLOGY OF MOBILE IP

MOBILE NODE. A mobile node is a host or router that changes its point of attachment from one network or subnetwork to another. A mobile node may change its location without changing its IP address. A long-term IP address is on a home network. When away from its home network, a care-of address is associated with it.

HOME AGENT. A home agent is a router on a mobile node's home network that tunnels datagrams for delivery to the mobile node when it is away from home. The tunnel is the path followed by a datagram while it is encapsulated. The datagram, then, is protected from normal Internet routing until it reaches a knowledgeable decapsulating agent. Tunneling is an action that bypasses the normal Internet routing of a packet by enclosing (encapsulating) the packet within a new IP header that contains an alternate destination IP address.

FOREIGN AGENT. A foreign agent is a router on a mobile nodes' visited network that provides routing services to the mobile node while it is registered. The foreign agent detunnels and delivers datagrams to the mobile node that were tunneled by the mobile node's home agent.

8.4.5 ROLE OF THE IETF FOR MOBILE IP

Mobile IP has been standardized through the mobile IP working group within the Internet Engineering Task Force (IETF), which is a collection of approximately 70 working groups. The deployment of mobile IP should probably proceed in two stages. The first stage is the base protocol, which allows for operation with no changes to existing Internet computers but suffers from suboptimal routing. The next stage is the route optimization, which finds the best ways to modify existing computers and better routers for mobile nodes. The firewall issues also need to be solved.

8.4.6 THE PROGRESS OF MOBILE IP

The application of mobile networks is different from the wireline network and is the weakness of the Internet. When a mobile phone moves from one base station to another, or from one mobile switch to another, the handoff functionality assures the continuity of the voice connection. In the Internet, if a user terminal moves and its connection point to the Internet changes, the numerical IP address of the device also needs to change. This causes problems because an existing session would have to be dropped and a new session started from the new location. To overcome this problem, the mobile IP protocol allows the home network (see Fig. 8.9) to keep track of the mobile device as it moves within the network or to other networks. Mobile IP uses protocol tunneling to hide a mobile node's home address from intervening routers between its home network and its current location when away from home (Fig. 8.9). The mobile IP protocol allows movement between different types of networks, from a corporate network to an ISP network to a cellular network, as long as these networks are IP-based packet networks supporting the protocol. Mobile IP provides a powerful tool for public and private network integration for packet data services.

Recognizing this potential, a standards body within the U.S. Telecommunications Industry Association, TR45.6, is working on a mobile IP compatible specification for supporting packet data in cdmaOne and TDMA (IS-136) systems. GPRS provides an end-to-end system specification for offering packet data services in GSM systems. GPRS uses IP for communication between some of the internal nodes and within the Internet. The mobility management of the data devices

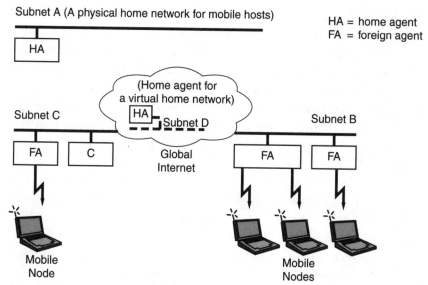

Figure 8.9.   Legend to come.

and application sessions is not compatible with mobile IP. Once the mobile IP protocol forms the basis for providing packet data services in mobile networks, it is conceivable that its use will be extended to support mobile voice applications as well. The mobile IP protocol is still evolving. However, commercial implementation of the current version of the standards already exists.

8.5   WIRELESS APPLICATION PROTOCOL (WAP)

8.5.1   INTRODUCTION

Phone.Com (formerly UnWired Planet) was a joint cofounder of the WAP Forum in June 1997, with Nokia, Ericsson, and Motorola. The Forum is primarily focused on a mass-market wireless phone as the means for delivery of Internet-based services. Because there will be 600 million wireless subscribers by 2001, this multimedia needs capabilities including the ability to retrieve email and push and pull information from the Internet. To guide the development of these exciting new applications, the WAP was developed for the presentation and delivery of wireless information and telephony services on mobile phones and other wireless terminals. WAP specifications address the above issues by using the best of existing standards and developing new extensions where needed. The WAP solution leverages the tremendous investment in Web servers, Web development tools, Web programmers, and Web applications, while solving the unique problems associated with the wireless domain. The specification further ensures that this solution is fast, reliable, and secure. The WAP specification extends and leverages existing technologies, such as digital data networking standards, and Internet technologies such as IP, HTTP, XML, SSL, uniform resource locators (URLs), and other content formats. For example, a WAP gateway is required to communicate with other Internet nodes using the standard HTTP1.1 protocol. The specification calls for wireless handsets to use the standard URL addressing scheme to request services.

8.5.2   INTERNET PROGRAMMING MODEL AND WAP PROGRAMMING MODEL

The Internet programming model is shown in Fig. 8.10a. The Web browser requests the service from Servlet at the Web server through an URL. The service delivery, using JAVA Script, responds through HTML to the Internet protocol stack at the client. The WAP programming model (see Fig. 8.10b) is heavily based on the existing Internet programming model. Introducing a gateway function provides a mechanism for optimizing and extending this model to match the characteristics of the wireless environment. Over-the-air traffic is minimized by binary encoding/decoding of Web pages and readapting the Internet protocol stack to accommodate the unique characteristics of a wireless medium such as call drops.

8.5.3   KEY COMPONENTS OF THE WAP TECHNOLOGY

1. *Wireless Mark-up Language (WML).*   Incorporates the concept of cards and decks, where a card is a single unit of

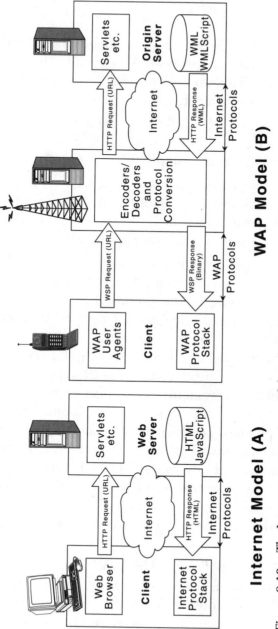

Internet Model (A)   **WAP Model (B)**

Figure 8.10. The Internet programming model vs. the WAP programming model.

interaction with the user. A service constitutes a number of cards collected in a deck. A card can be displayed on a small screen. WML Web pages reside on traditional Web servers.

2. *WML Script.*   A scripting language, like JAVA (see Sec. 8.6), that enables applets to be dynamically transmitted to the client device and allows the interaction with the user to be more sophisticated and intelligent.

3. *Microbrowser.*   An application resident on the wireless terminal that controls the user interface and interprets the WML/WMLScript content.

4. *Wireless Telephony Applications (WTA).*   A framework that provides a device-independent interface to allow network operators to enhance or build telephone services such as call control, phone book access, and messaging functions.

5. *A lightweight protocol stack.*   Minimizes bandwidth requirements, guaranteeing that a broad range of wireless networks (paging to the proposed 3G networks) can run WAP applications. The protocol stack of WAP comprises a set of protocols for the transport (WTP), session (WSP), and security (WTLS) layers. WSP is binary encoded and able to support header caching, thereby economizing on bandwidth requirements. WSP also compensates for high latency by allowing requests and responses to be handled asynchronously, sending before receiving the response to an earlier request. For lost data segments, perhaps due to fading or lack of coverage, WTP only retransmits lost segments using selective retransmission, thereby compensating for a less stable connection in wireless. The relationship of the functions in the operator's domain is shown in Fig. 8.11.

6. *Advantage of WAP.*   (1) The WAP protocol uses less than one-half the number of packets that the standard HTTP/TCP/IP Internet stack uses to deliver the same content. (2) Addressing the limited resources of the terminal, the browser, and the lightweight protocol stack are designed to make small claims on CPU and ROM. (3)

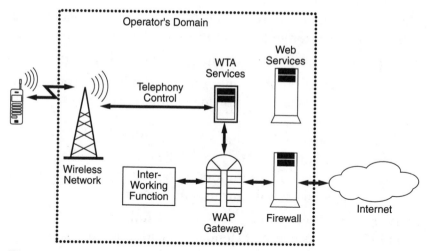

Figure 8.11.   Relationship of functions in the operator's domain.

Binary encoding of WML and WMLScript helps keep the RAM as small as possible. (4) Keeping the bearer utilization low takes account of the limited battery power of the terminal.

8.6   BLUETOOTH AND JINI

Bluetooth is a short-range radio that is used to connect electronic devices, such as PCs, together. Jini is a new system architecture that provides mechanisms for service construction, loop up, and communication and is used in a distributed system.

8.6.1   BLUETOOTH

INTRODUCTION.   Bluetooth is a radio unit that enables electronic devices to connect and communicate among themselves wirelessly via a short-range, ad hoc network. Bluetooth was first initiated by Ericsson, Nokia, IBM, Toshiba, and Intel. Now, more than 1000 companies support this unit. The name Bluetooth comes from the nickname of a Viking leader. Because mobility has become very important in our daily life, it is very attractive to use low-cost and easy-to-connect radios to bring the electronic devices together in an ad hoc network.

THE CHARACTERISTICS OF BLUETOOTH.   Bluetooth is a radio unit for short-range, point-to-multipoint voice and data transfer and has the following characteristics:

1. *Frequency band.*   Use of a 2.45-GHz ISM band, which is a license-free band open to communications as well as appliances, such as baby monitors, garage door openers, cordless phones, microwave ovens, and so forth.

2. *Range of transmission.*   The normal link range is from 10 cm to 10 m. It can be extended to 100 m by increasing the transmission power.

3. Bandwidth of each channel is 1 MHz.

4. Modulation is GFSK.

5. Adopts frequency hopping (FH), which divides the ISM frequency band into a number of hop channels.

6. Provides a system solution by specifying all seven layers of the OSI model. The Bluetooth architecture is shown in Fig. 8.12.

7. It is a master-slave-oriented architecture.

8. It supports up to eight devices in a piconet (i.e., two or more Bluetooth units sharing a channel).

9. It has built-in security.

APPLICATION OF BLUETOOTH.   Bluetooth is a system solution. It can work in a non-line-of-sight condition through walls and briefcases. It can easily integrate with other networks through TCP/IP. Bluetooth will enable users to connect to a wide range of computing and telecommunication devices and services without the need to buy, carry, or connect cables. It delivers opportunities for rapid ad hoc connections and, in the future, possibly for automatic, unconscious, connections between devices and services. Bluetooth's power-efficient radio technology can be used in many of the same devices that

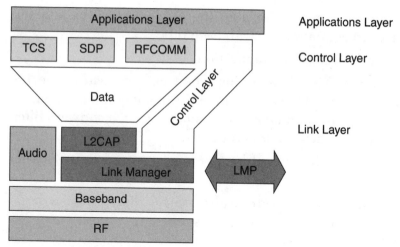

Figure 8.12.   Bluetooth architecture.

use IR: phones and pagers, modems, LAN access devices, headsets, notebook, desktop, and handheld computers.

8.6.2   JINI

DESCRIPTION.   The size of computers has become smaller, and the speed of computation has increased dramatically. Jini became Sun Microsystem's vision of the future of communications in the computer world. JAVA runs on all platforms. Jini addresses the critical needs for computer access to network services through a simple and uniform interface. Jini is a new system architecture that brings to the network the facilities of distributed computing, network-based services, seamless expansion, reliable smart devices, and ease of administration. The dynamic nature of a Jini system enables services to be added or withdrawn from a federation at any time according to demand, need, or the changing requirements of the working group who uses it.

CHARACTERISTICS.   Jini has the following characteristics:

1. Uses 100 percent JAVA language, developed by Sun. JAVA is described in Sec. 8.6.4.

2. Provides a means for allowing different devices and services to communicate with each other.

3. Is a robust network application.

4. A "service" is the most important concept within the Jini architecture. A service is an entity, a computation, storage, a communication channel to another user, a software filter, a hardware device, or another user.

5. A service can be used by a person, a program, or another service.

6. A Jini network system allows the distributed virtual machine to work as a single system.

KEY ELEMENTS OF JINI.   Two elements, discovery and lookup, in the architecture allow the devices and software to connect to and register with the network, as shown in Fig. 8.13:

1. Discovery and Join solve the difficult problem of how a device or application registers itself with the network for the first time with no prior knowledge of the network.

2. Lookup can be thought of as a bulletin board for all services on the network.

3. Communication between services can be accomplished using JAVA Remote Method Invocation (RMI). RMI is a JAVA programming language-enabled extension to a tradi-

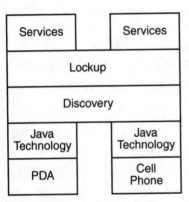

Figure 8.13.   Jini architecture.

tional remote procedure call mechanism. RMI allows data and full objects, including code, to be passed from objects around the network.

8.6.3 RELATIONSHIPS BETWEEN BLUETOOTH AND JINI

Bluetooth is a device and Jini is a service. The differences are shown in Table 8.1. Bluetooth is a system solution that defines the seven layers of the OSI model to establish connections and provide services. Jini is 100 percent based on JAVA and only provides the network service lookup and the discovery function for the existing network. Therefore, Bluetooth can enhance Jini by providing wireless networking capability.

8.6.4 JAVA BACKGROUND

BACKGROUND. JAVA is an interpretive language that requires an interpreter, the virtual machine (VM), to translate the software code into byte codes for the specific operating

Table 8.1

BLUETOOTH	JINI
Every *device* has an instance.	Every *service* has an instance of Jini.
Kernel and services are language independent.	Jini requires a homogeneous JAVA environment.
Encourages small and efficient application footprints.	JAVA applications and JVMs are large and slow (today).
No piano kernel functions rely on RMI device discovery.	All components communicate with RMI bandwidth-intensive Jini.
Device discovery.	Assumes preexisting network.
Service registration.	Discovery/Join.
Service discovery.	Lookup.
Client.	Proxy.
Kernel = 37KB	PJAVA JVM = 1MB
RTOS = ?	OS/9 = 1MB

system. Because the VM is device specific, JAVA is device independent, hence connotating the phrase "write once, run anywhere." JAVA is an open standard and is widely accepted in the industry.

There are three editions of the JAVA 2 software platform: enterprise edition, which is for servers; standard edition, which is for desktops; and micro edition, which is for handhelds, PDAs, embedded systems, and JAVA Cards (J2ME). Sun's virtual machine for J2ME, kVM, is as small as 128KB on the device and has functionality split between the device and external service providers.

DEVELOPMENTS.   NTT DoCoMo plans to incorporate JAVA into its i-Mode phones by the fall of 2000 before the introduction of IMT-2000. With the smaller footprint of kVM and the continually increasing CPU power and improving power consumption, JAVA will also likely be suitable for handsets.

Smart messaging and WAP are meant for "lite" phones, whereas JAVA aims at more sophisticated network terminals that run operating systems such as EPOC from Symbian and Windows CE from Microsoft. Although phone display sizes and computing power are increasing, they are still a limiting factor when it comes to running complicated JAVA applications.

The WAP Forum has a working group that is looking at incorporating JAVA into the WAP specification. It is likely that it will complement the mark-up language WML (it does not complement HTML) for the more dynamic browser applications that run over the WAP lightweight protocol stack.

Comparing some of the attributes of WAP (Fig. 8.10) and JAVA (Fig. 8.12) illuminates some of the advantages of incorporating JAVA into the phone.

8.7   WIRELESS IP CORE NETWORK

Wireless communications are growing so rapidly that in 2000 there are 600 million mobile handset users. By 2003, users will exceed 1 billion. At the same time, the world of Internet users also is growing rapidly. In 2000, there are more than 200 million Internet users.

The future is moving into wireless Internet services. Today, a cellular or PCS user can have a handset with microbrowser equipment to access an ISP. Today's network is shown in Fig. 8.14; the ISP is connected to the PSTN through a MTSO that is a circuit switch. It is very slow and inefficient to use the current network. In 2000, the packet switch will be installed in the GSM system of some European countries. The IP protocol from SGSN to GGSN and to the wire-line IP network is shown in Fig. 8.15. However, it is very costly and not a total IP solution. Two approaches to the wireless IP Core Networks have been proposed: the ATM Centric IP Core Network (shown in Fig. 8.16) and the Router Centric IP Core Network (shown in Fig. 8.17). The ATM Centric IP Core Network has quite an advantage, but the cost is relatively high. The advantages are as follows:

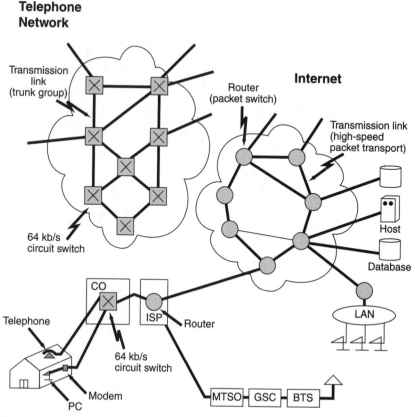

Figure 8.14. Telecommunications today.

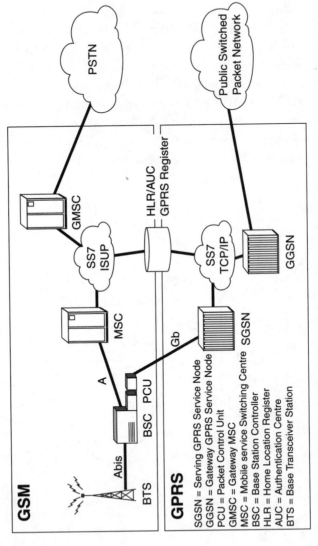

Figure 8.15. Today's GPRS architecture.

GSM

PSTN

GMSC

SS7
ISUP

MSC

A

BTS   BSC   PCU

Abis

GPRS

Public Switched
Packet Network

HLR/AUC
GPRS Register

SS7
TCP/IP

GGSN

Gb

SGSN

SGSN = Serving GPRS Service Node
GGSN = Gateway GPRS Service Node
PCU = Packet Control Unit
GMSC = Gateway MSC
MSC = Mobile service Switching Centre
BSC = Base Station Controller
HLR = Home Location Register
AUC = Authentication Centre
BTS = Base Transceiver Station

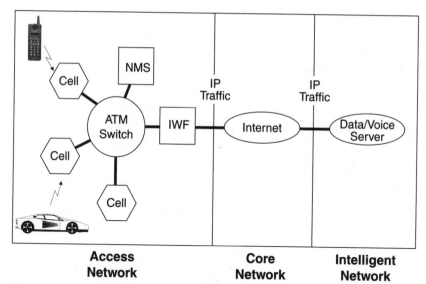

Figure 8.16.   ATM-based wireless IP network.

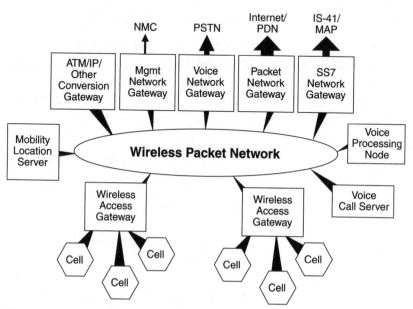

Figure 8.17.   Router-based wireless IP network.

1. The broadband switch
2. Has QoS
3. Has all the mobile switching functions
4. A centralization configuration approach

The disadvantages of ATM Centric approach are as follows:

1. Not a total wireless IP Core Network
2. High cost
3. Not easy to add on features and functions

The Router Centric approach (see Fig. 8.16) is a low-cost, simple-configuration, totally wireless IP core network solution. On December 15, 1999, four companies, Vodafone AirTouch, Cisco, Hyundai, and Telos, led by Vodafone AirTouch, held a demonstration of the wireless IP core network shown in Fig. 8.18 (see also Exhibit 8.A). W. Lee had his motivations for this demo. The motivations were

1. IP is the answer for multiservice networks that include next-generation voice and data wireless networks.
2. IP is gathering massive market and technological innovation momentum. The number of users and devices attached to IP networks has grown exponentially during the last few years. The amount of data traffic has surpassed the amount of voice traffic in many operator environments.
3. Vodafone AirTouch, Disco, Hyundai, and Telos have a common interest in developing a new wireless IP network.
4. A proof-of-concept demonstration using IP for a wireless system is critical to advance the wireless IP architecture.

The project goal was to demonstrate the technological and economic viability of a wireless IP network and the feasibility of integration with other IP-based wireless networks. The evaluation of the wireless IP network viability cover the following areas:

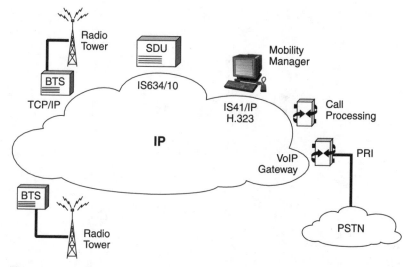

Figure 8.18.   Top-level architecture of wireless IP core network.

1. *Investment efficiency.*   Assess the trade-off model of an initial ongoing equipment investment, area coverage, network and system capacity, and system scalability.

2. *Operation efficiency.*   Study the benefits of common IP transport, network deployment, service provisioning, network availability, and packet network dimensioning.

3. *Network performance.*   Explore recent advances in QoS technology, explore the security aspect of packet airlink, and understand the challenges of meeting telecom requirements.

4. *Customer care efficiency.*   Evaluate the advantages of IP common interfaces, easy-to-integrate network operation with customer care, timely and multivendor customer care solution, e-bill, and e-help (i.e., e-business).

5. *Mobility essentials.*   Realize the implementation of mobility functions, which include mobility management, soft handoff, and radio resource management.

6. *Unbundled IP-based wireless network.*   Investigate the potential of plug and play with multivendor network elements, applications, and services.

Exhibit 8.A.   A demonstration for a wireless IP network solution, sponsored by Cisco, Hyundai, Telos, and Vodafone Airtouch.

7. *Is IP ready for telecom core element?*   Reveal deficiencies, if any, and suggest improvements. Investigate the advantages of an end-to-end IP network. Investigate the boundary capacity analysis for combined data and voice traffic on a single IP network.

8. *Industry influence.*   Define and standardize IP-based radio access networks and core network architectures, interfaces, and protocols. Form a strategy on necessary changes to the architecture of the wireless IP network. After the demo, Cisco and Vodafone Airtouch found the need and initiated Mobile Wireless Internet Forum (MWIF), in February 2000, to make recommendations to the 3G standard body in the wireless IP network area.

The demonstration features were as follows:

1. Presented a wireless end-to-end IP network architecture using the IS-95 CDMA system in Reno, Nevada.

2. Three types of calls were demonstrated through the wireless IP network, land to mobile calls, mobile to land calls, and mobile to mobile calls.

3. Demonstrated the handoff features in the IP Core network.

This was the first wireless IP network solution demonstration in the industry and a key step toward the success of a wireless IP network. It showed the ease of connecting network elements, while delivering seamless services and network management.
    The demonstration had the following unique features:

1. Elimination of switches. Cisco's Call Agent (CA) is the soft switch.

2. Total wireless IP network.

3. Separation of mobility management and call process, which enables system scalability and engages faster and cheaper subscriber feature development.

4. Separation of signaling and data.

5. Enhanced the potential of wire-line and mobile convergence.

6. Data network integrated with mobile network.

7. IS634 to H.323 conversion.

8.7.1 FUTURE IMPROVEMENTS

Although IP will be the core for connecting network elements and providing services, there are other areas that need to be improved:

1. QoS.   Several proposed technologies for a wireless internet network should be tested.

2. *Billing.*   The IP billing paradigm is different from the traditional "charge by minute" model. The ease and adaptability of the billing system for "charge by packet or Mbytes" needs to be investigated.

3. *Latency.*   Dimension the IP network so that time-critical data can be delivered on time.

8.8   INTERFERENCE OR NOISE?

In the past, the FCC has allocated a particular spectrum and license to a specified service provider. If the modulation scheme or the power emission used in the operated spectrum is not within the FCC's specifications, interference will rise outside the band and will be detected. The FCC could then discipline those who cause interference if complaints are made by the neighboring spectrum users to the FCC. The FCC's charter is to coordinate the spectrum utilization. Then the issues of spectrum management and electromagnetic noise are addressed.

8.8.1   SPECTRUM-SHARING POLICY

To achieve the objective of high-spectrum efficiency in utilization, the FCC is heading to a spectrum-sharing policy. There are three different scenarios in spectrum sharing.

MANY AUTHORIZED OPERATORS PROVIDE THE SAME SERVICE.   Today, the FCC is trying to utilize spectrum efficiently while it also creates fair competition in the market at the same time. Actually, spectrum efficiency is inversely proportional to the number of system operators.[12]

spectrum efficiency = (number of system operators)$^{-1}$

The above expression means that the more that system operators use the subdivided spectrum in the total allocated spectrum, the less efficient the spectrum. But to have fair competition, the highest allowable number of operators may have to be chosen in a shared spectrum bandwidth, based on the market demand and the tolerance of the interference.

MANY AUTHORIZED OPERATORS PROVIDE DIFFER-ENT SERVICES.   Today, the FCC wants to have two or more different services share in the same band. In fact, the benefit of the spectrum-sharing concept can hardly be realized. In military or public safety agent operations, the usage of their allocated spectrum is light. However, we cannot take a chance and affect the spectrum because each military call or emergency call could be related to life or death. In commercial use, there are three concerns to be addressed to demonstrate the weakness of the spectrum-sharing concept:

1. The market is continuously growing, but the allocation of spectrum from the FCC remains unchanged in the cellular and PCS spectrum. Supposedly, the cellular spectrum can be shared with other services today. What about tomorrow? If there is no limit to the number of different services, eventually no one can operate in the same spectrum.

2. The cellular and PCS spectrum will provide the E911 calls with their mobile locations. The cellular and PCS spectrum will be treated to this extent as a public safety spectrum, which cannot be affected by sharing it with other services.

3. The third concern is the CDMA signal transmission. It spreads its energy wide enough to be a noiselike signal. In CDMA systems, the noiselike signal can defeat the foreign interference coming from the operator of a different service, but it has to pay the price of sacrificing its radio's capacity. The spectrum-sharing concept actually is to transfer the spectrum usage from one service to another. No spectrum efficiency is achieved.

In the licensed cellular and PCS band, operators are planning to change from 2G to 2.5G to 3G systems. Operators are trying hard to use the spectrum efficiently. There is no room for foreign spectrum-sharing services in this band.

UNAUTHORIZED OPERATORS IN DIFFERENT SERVICES. Ultrabroadband transmission can use either the spread spectrum (SS) technique or pulse position modulation (PPM) to spread the signal over a spectrum bandwidth of 2 to 6 GHz. As a result, a noticeable source becomes an unnoticeable source:

signal transmitting from the operator

$$\text{intelligent signal} \xrightarrow{\text{becomes}} \text{undercover signal}$$

signal received by the other operators

$$\text{identified interference} \xrightarrow{\text{becomes}} \text{unidentified noise raised}$$

It is an analogy as

$$\text{a regular army} \xrightarrow{\text{turns into}} \text{a guerrilla force}$$
$$\text{(noticeable source)} \quad\quad \text{(unnoticeable source)}$$

In this situation, the general noise floor will keep rising, but the source will be hard to find and no one will take the blame for this. By allowing unauthorized operators to use the ultrabroadband spectrum, of course, the spectrum efficiency with an unlimited number of unauthorized operators can increase to the maximum, but the voice quality or the data transmission can be totally unacceptable.

At sending end, maximum spectrum efficiency means everyone can send calls without blocking and can send as many as desired. At receiving end, maximum spectrum efficiency means nobody will be able to understand or detect a received call.

8.8.2 Spectrum Efficiency and Noiselike Signal

Every problem can be viewed from at least two sides. From an individual's perspective, as long as the wide bandwidth spectrum is used, an individual may claim that his or her operation signal only contributes an insignificant noise level across the entire band. The FCC cannot detect the signal, so he or she can transmit as often as possible. No license is needed and the expensive license fee is saved.

From the FCC's perspective,

1. An ultrabroadband signal does not need a license from the FCC to operate.
2. Spectrum efficiency is achieved.
3. It will make the FCC's job more difficult to clean up the mess of high noise level in the future.
4. The auctioned licensees assume their ownership of the spectrum and will ask the FCC to stop any trespass signals through their spectrum.

8.8.3 Conclusion

The spectrum-sharing policy needs to be studied very carefully by the FCC. Otherwise, the wireless information age will be a dark age due to the high noise level.

8.9 WILL WIRELESS COMMUNICATION COME TO AN END?[13]

8.9.1 Introduction

In 1897, Marconi successfully set up a wireless communication link that was 18 miles long and ran from Needles, on the Isle of Wright, to a tugboat. Today, wireless communication has become a part of our daily lives. It is, therefore, appropriate to address the trend of wireless communication in the future.

During the past 30 years, the wireless communication industry has been growing very rapidly. Pagers, cordless phones, satellite communications, cellular phones, and PCS phones are very popular. Among these services, voice has always been the main focus for customers. Even when responding to a page, a phone device is used. However, in the future due to the growth of the Internet, data service will become important.

Digital transmission, including voice, data, computing, and entertainment, rely on high-speed data transmission. But, high-speed data transmission needs a wide bandwidth, and in the wireless communication medium, bandwidth is a precious commodity. It is almost impossible to receive wireless bandwidth as huge as the optical fiber bandwidth. Therefore, the question is raised, Will wireless communication come to an end? Before we answer this question, we must examine some key factors and problems. If we can find solutions to these problems, the answer will be clear.

The major concerns regarding the future of wireless communication are Mother Nature's limitations and human-made factors. Mother Nature's limitations consist of limited natural spectrums, demand and capacity issues, technology efforts, and intelligent systems. Human-made factors consist of service creation and the government's role, as stated in the following sections.

8.9.2 LIMITED NATURAL SPECTRUM

First, the spectrum of electromagnetic waves is a limited natural resource. Therefore, the efficient use of spectrum is a big challenge. In wireless communication, because of radio interference there is only a limited number of service systems that can operate within their own allocated spectrum and discriminate interference from neighboring spectrums.

Currently, many systems efficiently use spectrum by using SS modulation. SS modulation spreads signal energy across the wideband spectrum. It doesn't cause interference, but it does raise the noise floor. If all SS systems share the same wideband spectrum, the noise floor will rise to a level where no one system can operate. Fortunately, there is natural evidence that the data transmission rate can be higher when the link distance becomes shorter within a given spectrum bandwidth. We will apply this natural evidence to the future.

8.9.3 DEMAND AND CAPACITY ISSUES

We have found that voice quality and system performance, including data transmission, are inversely proportional to the service demand and system capacity.

(Voice quality and system performance) = (service demand

and system capacity)$^{-1}$

There is a great challenge for system operators to find new technologies to raise the bar so that voice quality can be maintained with increasing service demand. We need wideband channels for capacity and high-speed data transmission. Therefore, wide bandwidth is a key challenge for wireless communication in the future.

The future trend in wireless communication is depicted in Figure 8.19. There are two processes, revolution and evolution. Today, the narrowband system going to the wideband system and the low-capacity system going to the high-capacity system are in the revolutionary process. Other existing systems can go through the evolutionary process, such as the narrowband system to achieve high capacity and a low-capacity system to achieve wideband services. The G3G system is currently going from a narrowband, low-capacity system to a wideband, high-capacity system through a revolutionary process.

8.9.4 TECHNOLOGY EFFORT

Technology efforts should be directed toward finding ways to utilize EM spectrum efficiently.

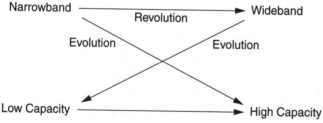

Figure 8.19.   Future trends in wireless communications.

WIDEBAND RADIO. Wideband radio is needed for high-speed data transmission. There are two kinds of wideband radios, the hardware radio and the software radio. The software radio (see Section 7.4) can be used more flexibly than the hardware radio in the following applications:

1. One wideband radio for one service, such as cellular or PCS, can provide wideband (WB) or narrowband (NB) channels, or mixed WB and NB channels.

2. One wideband radio for multiservices (such as cellular, PCS, satellite, etc.). In this case, the radio can serve one service at one time or serve multiservices at one time.

3. One wideband radio for multisystems (such as CDMA, GSM, AMPS, etc.).

4. One wideband radio for multiservices and multisystems. The radio must continue to maintain good voice quality after acquiring the desired signal through its intelligent search and must discriminate against all undesired signals. It is a challenge.

5. The technology of software radio is still in its research and development stage. The wideband software radio is hard to realize and needs a breakthrough technology.

MM WAVE AND IR. To apply wideband communications to fast data transmission, FAX, video, and so forth, wideband spectrum is needed. In wireline communication, the trend is changing from twin leads and coaxial cable to the wideband fiber cable. In the early 1970s fiber cable loss was very high. Technology drove the fiber cable loss down to 0.1 dB per kilometer, where it is today.

In wireless communication, the wideband spectrum can only be provided at higher frequencies. The frequency ranges have been moved up from HF, VHF, and UHF to microwave and infrared. However, the propagation loss of microwave and infrared communication today is very high, and their links must be in line-of-sight condition. In the future, we may find different ways to reduce the loss by deploying special kinds of reflectors in the field or using other apparatus. The same trend we went through in reducing the fiber cable loss would be followed in reducing the loss in microwave and infrared links.

SYSTEM CAPACITY. Within the allocated spectrum, maximizing system capacity is the other challenge. Today, spread spectrum modulation has proven to be a better way of increasing system capacity.

INTERFERENCE-FREE TECHNOLOGY. The sharp skirt filter, the broadband linear amplifier, and so forth, need to be improved to stop interference caused by the neighboring spectrum users. Superconductivity (low-temperature nature) technology is attempting to play a role in noise reduction and might be applied to filters and low-noise amplifiers.

HYBRID SYSTEM. The wideband communication system may have to be a hybrid system with fiber cable. The majority of the links in a communication network are fiber cables. The concerns are only the connection between the fiber cable and the radio link at the last 100 m. The last 100 m of wireless wideband technology may be much easier to develop and would be used for portable and mobile applications.

INTELLIGENT SYSTEMS. A good network structure for wireless networks is developed by (1) understanding the limitations of Mother Nature, (2) ingenuity from human-made effort, (3) the risk-taking software development. A system using the three parts shown in Fig. 8.20 may be called an intelligent system. The subscriber unit should be intelligent to execute orders. The cell sites have intelligence to pass on the orders, and the network has intelligence to make orders or commands.

8.9.5 NEW CONCEPTION OF SPECTRUM UTILIZATION

Since the wireless Internet is starting to grow fast, and spectrum auction price has become very expensive, wireless service providers must think of utilizing spectrum more consciously.

Figure 8.20. An intelligent network structure.

A. Spectrum acquisition becomes a very expensive resource.

- *The UK 3G Auction*—The United Kingdom, a country of 60 million in population, had a 3G spectrum auction of 140 MHz spectrum divided into five license bands. Among the five license biddings, the payment for license B band (2 × 15 MHz), was 6 billion UK pounds, or close to 10 billion U.S. dollars. The message derived from this auction is that the future of wireless mobile communications have to combine with the Internet to be a very prosperous business.

  Also, the cost of purchasing equipment relatively becomes unimportant as compared to the cost of auctioning the spectrum. Additionally, each service provider will be more cautious using spectrum in order to justify the auction payment.

- *The US 700 MHz Auction*—The United States will release TV channels 60 to 62 and 65 to 67 (six TV channels totaling 36 MHz) for wireless communication. Among them, 30 MHz will issue two licenses. One is 2 × 5 MHz (747–752, 777–782 MHz) and the other 2 × 10 MHz (752–762 MHz, 782–792 MHz). The 6 MHz are the guard band. The United States will be divided into six economic area groupings (EAGs). Each EAG area will have two licenses. The total licenses in the entire country are 12. The auction is planned to begin in September 2000. The prediction is that the U.S. government may obtain up to $20 billion of auction revenue for 30 MHz as predicted.

B. High-Speed Data

Because voice mixes with date in a 5-MHz 3G channel, the power for a 6-kbps voice traffic channel and the power for a 386 kbps data traffic channel are different. In a 5-MHz CDMA carrier, it is very difficult to handle two different kinds of traffic channels with existing power control scheme to solve the near-far interference problem while a data rate is higher than 386 kbps. Therefore, the efficiency of channel utilization in the 5-MHz carrier is low. In order to make the channel utilization high, a dedicated channel should be used to transmit a peak forward link data rate up to 2.4 Mbps called HDR (high data rate) developed by Qualcomm, who conducted a trial in San Diego. Additionally, other vendors also developed their high-speed data channels, 1Xtreme developed by Motorola and by Nokia, E-1XRTT by Nortel, and 1Xplus by Ericsson shown in Table 8.2.

C. Line-of-Sight Link Platform

In order to have high-speed data, we need to create line-of-sight (LOS) links. The LOS reduces the time delay spread, frequency spread, space spread, and angle spread while the signal is received through the medium. There are several ways to create the LOS links:

1. Last 100 meter links—These links can be LOS links and can be applied on millimeter wave or infrared for high bandwidth.

2. Broadband satellites—The GEO or LEO satellites can provide high-speed data, but cannot serve high capacity traffic mobile units. Their footprints are very large; therefore, the frequency reuse concept of the cellular system to increase spectrum efficiency is not feasible for the satellite systems with a limited bandwidth resource.

3. Atmospheric satellites—Atmospheric satellites are platforms that provide telecommunications services from an altitude in the stratosphere (~50,000 ft) and above commercial aircrafts' altitudes (~30,000 ft). Some are unmanned aerial vehicles. Others are special purpose aircraft. Each has a satellite-like telecommunication payload and an antenna system that radiates a pattern of beams onto the earth below. Typically, atmospheric satellites stay aloft for less than a day to several weeks or even years at a time. A single operations center with ground spares of aerial-vehicles can maintain 7-km × 24-km coverage of one or more cities from 1000 km away. Atmospheric satellites, which will become a commercial reality, attempt to provide cost-effective broadband connectivity to residences and businesses in a city using an architecture that promises ubiquitous coverage and high capacity on a city-by-city level.

Here are some examples of the equipment stated as follows:

a. Angel Technologies Co.—Uses altitude long endurance (HALO) manned aircraft

b. AeroVironment Co.—Uses long-endurance unmanned aerial vehicle

Table 8.2. Several high-speed data channel systems

	HDR*	1XTREME	E-1XRTT	1XPLUS
Proposed by	Qualcomm	Motorola, Nokia	Nortel	Ericsson
Peak FL Data Rate	2.4 Mbps	5.2 Mbps	3.7 Mbps	2.4 Mbps
Peak RL Data Rate	307.2 kbps	614.4 kbps	460.8 kbps	307.2 kbp
Average Throughput	~600 kbps	~400 kbps	~600 kbps	~600 kbps
FL Access Method	TDM (fat pipe)	CDM (shared pipe)	TDM (fat pipe)	TDM
FL Modulation	QPSK, 8-PSK, 16-QAM	QPSK, 8-PSK, 16-QAM, 64-QAM	FL: QPSK, 8-PSK, 16-QAM, 32-QAM	Same as HDR
RL Modulation	BPSK	BPSK	BPSK	BPSK
FL Power Control	None, use full power available	Closed loop @ 800 Hz	None, use full power available	Same as HDR
RL Power Control	Open/closed loop @ 800 Hz	Open/closed loop @ 800 Hz	Open/closed loop @ 800 Hz	Same as HDR
Voice Service Support	No	Yes	Yes if concurrent with data	Yes
Transmit Diversity	None specified	Space Time Block Coding	Open-loop and closed-loop array technologies	None specified
Net. Arch - RLP/MAC	Decentralized	Not discussed	Decentralized	Decentralized
Net. Arch - Mob. Mgmt.	Decentralized	Not discussed	Centralized	Centralized
Mobility	Mobile	Nomanic	Mobile	Mobile

*The system had a public trial in San Diego, in November 1999.

    c. SkyStation Co.—Uses unmanned lighter-than-air platform

D. Modulation Technologies for LOS Environment

    1. In the nonfading environment, the modulations can be selected differently. The SSB modulation for vice and orthogonal frequency division modulation (OFDM) for data are spectrum-efficient modulation for the LOS environment (see Section 3.11.3).

    2. AlphaCom technology[14] also can be used. It can send 48-kbps MD-3 data stream through a 2-kHz filter and receives with good quality. The idea is to find ways to slightly mark the states of the modulation and VMSK (variable phase shift keying) on the carrier wave such that less distortion of the carrier waveform can be achieved. Of course, we know that undistorted CW carrier only needs 1-Hz filter in principle.

Because of the high cost of aquiring 3G spectrum, the utilization of spectrum for maximum efficiency will be the future research. Lee has made a prediction in many international conferences.[15–17] He said that today TDD is one of three 3G modes. In 4G, TDD will be the only mode. This is because TDD is a high-spectrum efficient system (see Section 4.12.5). Of course, we need to find the breakthrough technology that will make this happen. A smart coding scheme can reduce interference. LinkAir uses Prof. Dauben Li's codes and has demonstrated the code's merit.[18] Some vendors say coding technology was discovered in the 1960s. It's an old technology. Lee recalls coding in 1989, when Qualcomm's CDMA was demonstrated. Vendors also say that CDMA technology and spread-spectrum, technology were introduced in the 1950s. They are old technologies. The original technology may be old, but the modification of the original technology to a new system is a huge contribution. If coding technology is not a sufficient enough advance for TDD, then other breakthrough technologies will be found. Lee had a vision on the CDMA system in 1989, now he has a vision on the TDD system for the future.

8.9.6   THE GOVERNMENT'S POLICY

SPECTRUM FLEXIBILITY.   Currently, the FCC has set a policy to have spectrum flexibility, which can be classified in two areas—service and technical flexibility. The purpose of this policy is to promote competition and public interest by using the spectrum more efficiently. Service flexibility provides more services in an allocated spectrum in the same geographical area. Technical flexibility provides different technology systems in the same spectrum band and the same geographical area. It does cause some concern. This flexibility policy can run into two situations—instability and lack of discipline. This needs further study.

WHAT WOULD BE A GOOD INTERNATIONAL STANDARD?   A good standard system should be used by many different services. For example, the four services, PCS, Cellular, LEO/MSS, and WLL use one system. It is a desirable solution. The undesirable situation is numerous systems serving the same service. Today, three systems are operating for the PCS service; CDMA, NATDMA, and DCS 1900. Also, a good system should serve the subscriber's interest as follows: low cost, high quality, small size, light weight, long talking time, calls can be received anywhere and any time, and the unit is easy to handle and operate.

8.9.7   WIRELESS INFORMATION SUPERHIGHWAY

The information superhighway concept came from Vice President Al Gore. The wire-line information superhighway can be realized because of the huge bandwidth gained from the fiber optic network. In here, the only human-made issue is the network access. The access scheme to get on and off the information superhighway. In wireless communications, the poor limited bandwidth comes from the limitation of Mother Nature. Also, mobility causes another difficulty in pursuing this high-speed data transmission requirement. To obtain a large bandwidth, the carrier frequencies must be up to millimeter wave or to the infrared spectrum region. Over the millimeter wave links, we can apply the diversity scheme to increase the signal strength. Over the infrared links we can use diffuse transmission to create multipaths and operate under out-of-sight conditions. Also, because the infrared can penetrate rainfall but not fog, and the millimeter wave can penetrate fog but not rainfall, we can create a dual-medium diversity receiver that uses both infrared and millimeter waves for the last 100-m wireless high-speed data link (see Section 5.12).

8.9.8 CONCLUSION

Many areas of concern have been mentioned. If the technology effort cannot be moved ahead, if the service creation strategy timing is incorrect, if government does not play a strong role in spectrum coordination, and if demand and capacity are not well thought out and planned, we may possibly face an end to advances in wireless communications.

Of course, by making everyone aware of the various problems that lie ahead and by addressing these problems early on, we will be able to correct our course as we move forward into a great wireless communication future.

8.10   REFERENCES

1. Daniel C. Lynch and Marshall T. Rose, *Internet System Handbook*, Addison-Wesley, New York, 1993.
2. David Sacks and Henry Stair, *Hand-On Internet*, Prentice Hall, Englewood Cliffs, N.J. 1994.
3. John S. Quarterman, *The Internet Connection*, Addison-Wesley, New York, 1994.
4. L. Goldberg, "Wireless LANs: Mobile Computing's Second Wave," *Electronic Design*, June 26, 1995.
5. J. Cheah, "A Proposed Architecture and Access Protocol Outline for the IEEE 802.11. Radio LAN Standards, Part II," *IEEE Documentation*, p. 802, 11/91/54,
6. Charles E. Perkins, *Mobile IP, Design Principles and Practices*, Addison-Wesley, New York, 1998.
7. WAP Forum, "Wireless Application Protocol," *Wireless Internet Today*, June 1999.
8. Andrew Seybold, "Bluetooth Technology: The convergence of communication and computing." Andrew Seybold's outlook, May 16, 2000, *www.bluetooth.com*.

9. Ken Arnold, et al., "The Jini Specification," The Jini™ Technology Series, June 1999.

10. Jeffrey C. Rice and Irving Salisbury, *Advanced Java 1.1 Programming* McGraw-Hill, New York, 1997.

11. W.C.Y. Lee, "A Wireless IP Network Solution," conference of Reno Demonstration on IP Core Network, Reno, NV, Dec. 15, 1999.

12. W.C.Y. Lee, "Interference or Noise?, FCC Has to Make a Choice," submitted to FCC Technology Advisory Council, May 14, 1999.

13. W.C.Y. Lee, "Will Wireless Communications Come to an End?" *Journal of Information Science and Engineering.* 15:643–651, September 1999.

14. H.R. Walker, J.C. Pliatsikas, Dr. C.S. Koukourlis, and Dr. J.N. Sahalos, "Wireless Communications Using Spectrally Efficient VMSK/2 Modulation" in *Third Genersration Mobile Systems*, Springer Verlag, Berlin, 2000.

15. W.C.Y. Lee, "G3G and Its Future," 9th Annual Wireless and Optical Communications Conference, April 14–15, 2000, Marriott Airport Hotel, Newark, New Jersey.

16. W.C.Y. Lee, "Status and Future Prospects for Mobile Phone and Data Communications in the U.S.A. ," IEEE VTC-2000 Conference–Panel PA-03 "Deployment of IMT-2000," Tokyo, Japan, May 18, 2000.

17. W.C.Y. Lee, "Introducing the New Tools and Techniques for 3G," Workshop #2, CDMA World Congress, Hong Kong, China, June 12, 2000.

18. LinkAir Communication Inc. "LinkAir Communications LAS-CDMA Technology Seminar" May 8, 2000, Excalibur Hotel, Los Vegas, Nevada.

INDEX

Note: Boldface numbers indicate illustrations.

ABOUT THE AUTHOR

William C.Y. Lee, Ph. D., serves as chairman, as well as Chairman of the Board of Directors, for LinkAir Communications, Inc., the developer of LAS-CDMA—a new patented technology for wireless telecom systems that significantly increases network capacity, quality, and coverage over existing digital standards.

Regarded as a world-class scholar in wireless communications, Dr. Lee is renowned for his leading contributions in making analog and CDMA technologies commercially viable. As part of this effort, he has published more than 200 articles and several books on CDMA theory and technology. An expert in developing marketable communications technologies, Dr. Lee also invented and patented a new microcell system that increased radio capacity by 2.5 times over the conventional microcell model. He holds more than 25 U.S. patents, with more pending.

For 15 years, Dr. Lee was one of the pioneers in developing the advanced wireless technology for Bell Labs. He then joined the ITT Defense Communications Division, where he headed the advanced mobile communications system. During his subsequent tenure with Vodafone Airtouch, the world's largest mobile telecommunications company, Dr. Lee was instrumental in conducting key CDMA research and the initial trial of the technology. A leader in personal communications network (PCN) technology, Dr. Lee led the team that won the PCN license in the United Kingdom in 1989 for Vodafone. Dr. Lee also headed the application of PacTel's PCS experimental trial in 1993, and under his leadership, the first CDMA commercial system was completed in Los Angeles in 1995.

Dr. Lee has been elected as an I.E.E.E. Fellow and has served as a member of numerous councils, including the California State Council on Science and Technology, the U.S. Council on Competitiveness and the F.C.C. Technical Advisory Council.

He has earned many prestigious industry awards, including the IEEE VTS Avant Garde Award, the C.T.I.A. Award, the CDMA Industry Achievement Award, the SATEC Award, a Bell Lab Service Award, and most recently, the I.E.E.E. Third Millennium Medal Award.

Dr. Lee earned his doctorate in electrical engineering from the Ohio State University.